OCEANOGRAPHY AND OCEAN ENGINEERING

THE UNITED STATES AS AN OCEAN NATION

SCIENTIFIC RESEARCH PRIORITIES AND PROGRESS

OCEANOGRAPHY AND OCEAN ENGINEERING

Additional books in this series can be found on Nova's website under the Series tab.

Additional E-books in this series can be found on Nova's website under the E-book tab.

ENVIRONMENTAL SCIENCE, ENGINEERING AND TECHNOLOGY

Additional books in this series can be found on Nova's website under the Series tab.

Additional E-books in this series can be found on Nova's website under the E-book tab.

OCEANOGRAPHY AND OCEAN ENGINEERING

THE UNITED STATES AS AN OCEAN NATION

SCIENTIFIC RESEARCH PRIORITIES AND PROGRESS

MIKKEL JORGENSEN
EDITOR

nova publishers
New York

For permission to use material from this book please contact us:
Telephone 631-231-7269; Fax 631-231-8175
Web Site: http://www.novapublishers.com

Library of Congress Cataloging-in-Publication Data

ISBN: 978-1-62618-705-4

Published by Nova Science Publishers, Inc. † New York

Contents

PREFACE

The environmental integrity and sustainable productivity of the ocean, our coasts and coastal watersheds, and the Great Lakes play a central role in the well-being of all Americans. Every day, the marine environment supplies a multitude of products and services that enhance and support the lives and livelihoods of citizens. More than half of the United States population lives in coastal watershed counties, and these counties generate 58 percent ($8.3 trillion) of the Nation's gross domestic product (GDP), even though they comprise only 25 percent of the Nation's land area. Structured around six societal themes, this book recommends research priorities designed to advance our understanding of critical ocean processes and phenomena that are relevant to human health, economic well-being, environmental sustainability, adaptation to climate and other environmental change, and national and homeland security.

Chapter 1 – The environmental integrity and sustainable productivity of the ocean, our coasts and coastal watersheds, and the Great Lakes play a central role in the well-being of all Americans. Every day, the marine environment supplies a multitude of products and services that enhance and support the lives and livelihoods of citizens. In 2011, Americans, on average, ate 15 pounds of fish and shellfish per person – 4.7 billion pounds all together – making the United States second in the world in total seafood consumption. Offshore oil production in Federal waters accounts for fully 24 percent of total U.S. crude oil production. More than half of the United States population lives in coastal watershed counties, and these counties generate 58 percent ($8.3 trillion) of the Nation's gross domestic product (GDP)—even though they comprise only 25 percent of the Nation's land area. If American coastal watershed counties were considered an individual country, that country would have a GDP higher than that of China.

Recognizing the central role these environments and their resources play, and the many sometimes-competing demands placed on them, President Obama issued Executive Order 13547, *Stewardship of the Ocean, Our Coasts, and the Great Lakes*, on July 19, 2010.[4] The Executive Order established a comprehensive National Ocean Policy to improve the Nation's stewardship of the ocean, our coasts, and the Great Lakes. The Policy, which adopts the *Final Recommendations of the Interagency Ocean Policy Task Force*,[5] calls for use of the best available science to inform decisions affecting these essential ecosystems as well as to enhance humanity's capacity to understand and respond to a changing global environment. Full implementation of the National Ocean Policy will require significant progress in ocean science and more effective application of research results to aid decision-making. These two important concepts drove the selection of issues that frame this document, *Science for an Ocean Nation.*

Science for an Ocean Nation sets priorities for scientific research that support the National Ocean Policy goal of balancing the productive use of the marine environment with its stewardship and protection. It represents the consensus view of 25 Federal agencies regarding key ocean science research areas and incorporates extensive input from outside the Federal Government. [6] This report is constructed around six societal themes that frame ocean research in terms of the needs of Americans. Within these societal themes, *Science for an Ocean Nation* presents research priorities with the greatest potential to provide benefits to the Nation.

This report builds on its predecessor, *Charting the Course for Ocean Science in the United States for the Next Decade ("Charting the Course")*,[7] by updating research priorities, addressing emerging areas of research, and highlighting accomplishments resulting from Charting the Course. It also highlights cross-cutting capabilities and approaches that can help provide the technological advancements, broad perspective, and community structure needed to address the research priorities.

The themes that frame *Science for an Ocean Nation* represent key areas of interaction between human society and the ocean. The research priorities within each societal theme draw upon both natural and social sciences and aim to improve the tools required not only to carry out research but also to translate, disseminate, and apply the results of that research to address pressing issues facing the Nation and the world.

Chapter 2 - The Executive Order, authored by President Barack Obama, released by the White House of the Press Secretary dated on July19, 2010.

In: The United States as an Ocean Nation
Editor: Mikkel Jorgensen
ISBN: 978-1-62618-705-4

Chapter 1

SCIENCE FOR AN OCEAN NATION: UPDATE OF THE OCEAN RESEARCH PRIORITIES PLAN*

Subcommittee on Ocean Science and Technology

Executive Office of the President
National Science and Technology Council
Washington, D.C. 20502

Dear Colleague:

I am pleased to transmit this document, *Science for an Ocean Nation: An Update of the Ocean Research Priorities Plan*, which presents national research priorities in key areas of interaction between society and the ocean.

Our Nation's first ocean research priorities plan, *Charting the Course for Ocean Sciences in the United States*, was developed during 2005-06 and published in January 2007. Although it remains useful for many purposes, a number of issues not thoroughly addressed in that report have since risen in importance, including ocean acidification and rapidly changing conditions in the Arctic Ocean. Further, in July 2010 President Obama signed Executive Order 13547 establishing America's first National Ocean Policy, which calls for science-based decision-making as the Nation works to manage our ocean

* The report was released by the National Science and Technology Council, February 2013.

resources. For these and other reasons, the interagency National Science and Technology Council has crafted this update.

Structured around six societal themes, this report recommends research priorities designed to advance our understanding of critical ocean processes and phenomena that are relevant to human health, economic well-being, environmental sustainability, adaptation to climate and other environmental change, and national and homeland security. The report also provides updates on research progress in a number of these areas.

I am confident that the full spectrum of stakeholders with interests in the health of the global ocean and the importance of the ocean to society will find this report to be of great value.

Sincerely,
John P. Holdren
Assistant to the President for Science and Technology Director,
Office of Science and Technology Policy

EXECUTIVE SUMMARY

The environmental integrity and sustainable productivity of the ocean, our coasts and coastal watersheds, and the Great Lakes play a central role in the well-being of all Americans. Every day, the marine environment supplies a multitude of products and services that enhance and support the lives and livelihoods of citizens. In 2011, Americans, on average, ate 15 pounds of fish and shellfish per person – 4.7 billion pounds all together – making the United States second in the world in total seafood consumption.[1] Offshore oil production in Federal waters accounts for fully 24 percent of total U.S. crude oil production. More than half of the United States population lives in coastal watershed counties, and these counties generate 58 percent ($8.3 trillion) of the Nation's gross domestic product (GDP)—even though they comprise only 25 percent of the Nation's land area.[2] If American coastal watershed counties were considered an individual country, that country would have a GDP higher than that of China.[3]

Recognizing the central role these environments and their resources play, and the many sometimes-competing demands placed on them, President Obama issued Executive Order 13547, *Stewardship of the Ocean, Our Coasts, and the Great Lakes*, on July 19, 2010.[4] The Executive Order established a comprehensive National Ocean Policy to improve the Nation's stewardship of the ocean, our coasts, and the Great Lakes. The Policy, which adopts the *Final Recommendations of the Interagency Ocean Policy Task Force*,[5] calls for use of the best available science to inform decisions affecting these essential ecosystems as well as to enhance humanity's capacity to understand and respond to a changing global environment. Full implementation of the National Ocean Policy will require significant progress in ocean science and more effective application of research results to aid decision-making. These two important concepts drove the selection of issues that frame this document, *Science for an Ocean Nation.*

Science for an Ocean Nation sets priorities for scientific research that support the National Ocean Policy goal of balancing the productive use of the marine environment with its stewardship and protection. It represents the consensus view of 25 Federal agencies regarding key ocean science research areas and incorporates extensive input from outside the Federal Government. [6] This report is constructed around six societal themes that frame ocean research in terms of the needs of Americans. Within these societal themes, *Science for an Ocean Nation* presents research priorities with the greatest potential to provide benefits to the Nation.

This report builds on its predecessor, *Charting the Course for Ocean Science in the United States for the Next Decade ("Charting the Course")*,[7] by updating research priorities, addressing emerging areas of research, and highlighting accomplishments resulting from Charting the Course. It also highlights cross-cutting capabilities and approaches that can help provide the technological advancements, broad perspective, and community structure needed to address the research priorities.

The themes that frame *Science for an Ocean Nation* represent key areas of interaction between human society and the ocean. The research priorities within each societal theme draw upon both natural and social sciences and aim to improve the tools required not only to carry out research but also to translate, disseminate, and apply the results of that research to address pressing issues facing the Nation and the world.

Theme 1: Stewardship of Natural and Cultural Ocean Resources

Stewardship is at the core of the National Ocean Policy. The Policy calls for ecosystem-based management (EBM) as a guiding principle and approach, and coastal and marine spatial planning (marine planning) as a tool for achieving it. Ecosystem-based management requires intricate knowledge of ecosystems and consideration of the interactions within and between them. It also recognizes humans as integral parts of the equation. Research priorities within this theme focus on understanding changes in resource abundance and distribution, habitat–species and interspecies relationships, and interactions among human use patterns, resource stability, and sustainability. They also aim to enhance the societal benefits of ocean, coastal, and Great Lakes natural resources.

Priorities:

- Understand the status and trends of resource abundance and distribution through more accurate, timely, and synoptic assessments;
- Understand interspecies and habitat–species relationships to support forecasting of resource stability and sustainability;
- Understand human-use patterns that may influence resource stability and sustainability; and
- Conduct applied research using advanced understanding to enhance the benefits of various natural resources from the ocean, coasts, coastal watersheds, and Great Lakes.

Theme 2: Increasing Resilience to Natural Hazards and Environmental Disasters

Impacts to coastal and Great Lakes ecosystems and communities from natural hazards and environmental disasters range from disruption of processes and services to destruction of homes and habitats. Assessments of risk, vulnerability, and resilience are necessary to fully implement ecosystem-based management and restoration, planning, and recovery activities, and are required elements for marine planning. Development in areas at risk (e.g., from wind and floods) is the most significant contributor to the impact on humans of natural disasters, and it is likely that climate change will cause an increase in losses.

The amplified changes taking place in the Arctic are of special concern. Research priorities under this theme focus on understanding and forecasting natural hazards; assessing community and ecosystem risks; and supporting development of models and other decision-support tools, policies, and strategies for hazard mitigation.

Priorities:

- Understand how hazard events initiate and evolve, and apply that understanding to improve forecasts of hazard events;
- Understand the response of coastal and marine systems to natural hazards and apply that understanding to assessments of future vulnerability to natural hazards, including the increased vulnerability caused by climate change; and
- Develop multi-hazard risk assessments and support development of models, policies, and strategies for hazard mitigation.

Theme 3: Maritime Operations and the Marine Environment

Recreational boating, commercial and recreational fishing, shipping, military operations, and energy exploration and development provide critical benefits to society while also having impacts on ocean, coastal, and Great Lakes environments. At the same time, fleets of ships and other marine systems serve as platforms for scientific observation that can return data on changing conditions worldwide. Priorities under this theme focus on understanding the interactions between maritime operations and the environment, and using that understanding to predict changing conditions in the marine environment.

Priorities:

- Understand the interactions between marine operations and the environment; and
- Improve understanding of environmental factors affecting marine operations to better characterize and predict conditions in the maritime domain.

Theme 4: The Ocean's Role in Climate

The National Ocean Policy emphasizes the importance of better integrating ocean and coastal science with the broader domain of climate observations and climate modeling to assess vulnerabilities, develop adaptation strategies, and make policy and management decisions. Research priorities under this theme focus on understanding ocean–climate interactions within and across regions, understanding the impact of climate change and changing ocean biogeochemistry (including ocean acidification) on marine ecosystems, and projecting future climate change and its impacts, particularly in the Arctic. Such projections of changing conditions form the foundation for the adaptive management and planning efforts espoused by the National Ocean Policy.

Priorities:

- Understand ocean–climate interactions within and across regions;
- Understand the impact of climate variability and change on the biogeochemistry of the ocean and the implications for ocean ecosystems; and
- Apply understanding of the oceans to help project future global climate changes and their impacts.

Theme 5: Improving Ecosystem Health

The National Ocean Policy sets forth broad goals for ecosystem protection and restoration. Understanding ecosystems is a necessary step toward being able to improve their health. Data and research insights are needed to articulate requirements for ecosystem assessments and forecasts; address vulnerabilities and risks; increase resilience; and inform tradeoffs and priorities. The research priorities under this theme emphasize understanding and predicting the impact of natural and human-caused processes on ecosystems, including ocean acidification; developing socioeconomic assessments and models of human impact on ecosystems; developing appropriate metrics for sustainable management; and ensuring adequate restoration measures following major natural or human-induced catastrophes.

Priorities:

- Understand and predict the impact of natural and anthropogenic processes on ecosystems;
- Apply understanding of natural and anthropogenic processes to develop socioeconomic assessments and models to evaluate the impact of multiple human uses on ecosystems; and
- Apply understanding of marine ecosystems to develop appropriate indicators and metrics for effective management and sustainable use.

Theme 6: Enhancing Human Health

Human health is linked to the health of ocean, coastal, and Great Lakes environments. This theme includes priorities to use knowledge of ocean ecosystems and biodiversity to understand ocean processes that create risks or benefits to human health, as well as how ocean-borne human health threats affect the use and valuation of ocean resources. It also recognizes that decision making and coordinated action at local to global levels to protect human health requires science-based information, research on community adaptation and resilience, and increased public awareness of both concerns and opportunities.

Priorities:

- Understand sources and processes contributing to ocean-related risks to human health;
- Understand human health risks associated with the ocean and the potential benefits of ocean resources to human health;
- Understand how human use and valuation of ocean resources can be affected by ocean-borne human health threats and how human activities can influence these threats; and
- Apply understanding of ocean ecosystems and biodiversity to develop products and biological models to improve human well-being.

PREFACE

In 2007, the 25 Federal agencies represented on the National Science and Technology Council Subcommittee on Ocean Science and Technology (SOST – then named the Joint Subcommittee on Ocean Science and Technology) published *Charting the Course for Ocean Science in the United States for the Next Decade ("Charting the Course")*. This community-based national plan described what were, at that time, the most urgent ocean research issues at the intersection of society and the ocean. It pointed to areas in need of attention to better understand the ocean and its resources and to shape decisions regarding their use and protection.

In the six years since *Charting the Course* was released, much has changed. Most significantly, a National Ocean Policy and associated governance structure have been created for ocean, coastal, and Great Lakes stewardship. Furthermore, significant scientific progress has led to new insights, and new research areas have emerged. In this document, the SOST presents updated ocean research priorities that reflect events and accomplishments that have occurred since *Charting the Course* was published. It more fully incorporates what were, in 2007, newly emerging issues, such as changing ocean chemistry and decreasing ice cover in the Arctic related to global change. *Science for an Ocean Nation* builds on new knowledge of the ocean, our coasts, and the Great Lakes, and improvements in our ability to gain access to these environments. It acknowledges the growing importance of human-caused and natural disasters, which make clear the need for sound science to inform difficult decisions regarding the use and protection of resources. This document also more prominently recognizes the important role of social and behavioral sciences in decision-making and in implementing change to better understand and manage ocean, coastal, coastal watershed, and Great Lakes resources.

Science for an Ocean Nation is informed by our current state of knowledge, but also looks ahead to a future in which our understanding of the ocean and our interactions with it reaches new heights. This updated plan presents shared agency priorities to achieve that future – one endowed with the robust scientific foundation needed to guide society's wise use, stewardship, and enjoyment of the ocean and ocean-related resources.

Photo 1. Kelp and sardines, Anacapa Island, Channel Islands National Marine Sanctuary, California. Credit: National Ocean Service, National Oceanic and Atmospheric Administration.

INTRODUCTION

America has been a maritime nation since its inception, when shipping and commerce, maritime security, and fishing were critical to our fledgling country. As our country and population have grown, maritime commerce and ocean and coastal resources have become more important to our national prosperity, while ocean and coastal hazards and disasters have become more damaging. In recognition of the ocean's vast wealth of resources, vulnerable health, and our responsibility to be informed stewards, Presidential Executive Order 13547, *Stewardship of the Ocean, Our Coasts, and the Great Lakes*, was issued on July 19, 2010, adopting the *Final Recommendations of the Interagency Ocean Policy Task Force*[8] and providing for a comprehensive and integrated National Ocean Policy designed to improve the Nation's stewardship of the ocean, our coasts, and the Great Lakes. A unifying theme throughout the National Ocean Policy is that responsible use and stewardship of resources should be informed by and consistent with the best available science.

Photo 2. Lewes and Rehoboth Canal, Delaware. Credit: National Ocean Service, National Oceanic and Atmospheric Administration.

Together, *Science for an Ocean Nation* and the National Ocean Policy seek to dramatically change the way we understand, use, manage, and protect the ocean, our coasts and coastal watersheds, and the Great Lakes. *Science for an Ocean Nation* establishes priorities to guide science that supports the National Ocean Policy's goal of balancing the use of the marine environment with its stewardship and protection. As a country, we already are making decisions about resource use, whether balancing oil exploration and fisheries, developing shipping strategies for the Arctic, or setting aside areas for conservation. In the future, we will face increasingly multifaceted choices regarding the fate of coastal cities threatened by rising sea level, the health of marine ecosystems that are changing due to climate and other influences, and the use of areas and resources to accommodate new and emerging industries and technologies. Moving forward, the decisions and trade-offs that accompany these choices likely will become more complex and more consequential to our Nation's security and prosperity. Coping with change of the magnitude envisioned requires both significant progress in ocean science and transformation of how ocean scientists connect with the larger world. These two requirements informed the societal themes and cross-cutting topics that underpin this document. Success in bringing about change will require new intellectual partnerships, particularly between natural scientists who can model environmental change, economists who can assess trade-off options, and social and behavioural scientists who can address motivations for and consequences of change. Effective decision making and adaptation will also require an engaged public and broader and stronger partnerships among scientists, policy makers, and managers at all levels, both domestically and abroad. Preparing for and addressing change will require significant improvement in ocean observations and models scaled to both societal and scientific needs. Meeting these needs will cause ocean science to move rapidly toward solutions, regardless of the theme or topic under consideration. These requirements for success are common to all the societal themes and are further developed in the discussion of cross-cutting topics.

Why Now?

Charting the Course, which was drafted during 2005 and 2006 and published in early 2007, was developed as a prioritized decadal plan for interdisciplinary research to better manage the ocean challenges known at the time and those expected over the next decade. It called for revisiting the plan on a regular basis. This report responds to that call by addressing emergent priorities and several events that have taken place since 2007. In March 2009, the Omnibus Public Land Management Act of 2009 (Public Law 111-11) was enacted, calling attention, again, to the importance of understanding and properly managing ocean, coastal, coastal watershed, and Great Lakes resources. The law includes five sections on critical ocean and coastal issues including exploration, observations, ocean acidification, and mapping, spurring further action in these areas. The legislation also reenlists the Subcommittee on Ocean Science & Technology (SOST) and its member agencies to work with the wider ocean community toward progress. In June 2009, the President established the Interagency Ocean Policy Task Force to develop our Nation's first National Ocean Policy. The National Ocean Policy strongly identifies the integral role of science in meeting its goals and specifically includes policy points, principles, and objectives that call for scientific understanding, education, and technological advancement. Science for an Ocean Nation links this science to societal needs and provides a framework that will allow ocean research priorities to be quickly and effectively implemented to advance the National Ocean Policy's objectives. The Deepwater BP oil spill highlighted the wide and reverberating impacts that the loss of ocean and coastal resources can have on our country, and the critical importance of effective stewardship. It also emphasized the complex interplay of society and human health with chemical, physical, and biological aspects of the ocean and our coasts. It revealed significant areas in need of improvement, including our ability to monitor and observe the ocean in real time; to rapidly share information; to understand, model, and predict the consequences of extreme events in the context of a changing ocean; and to apply those insights to a rapid and effective response.

Revisiting the Ocean Research Priorities

In updating priorities for ocean research, the SOST undertook considerable outreach to the ocean community as a whole (see Box 1). Input from the wider community and extensive internal discussion revealed that the six societal themes presented in Charting the Course remain timely and relevant today. Thus, they were retained as the basis of Science for an Ocean Nation:

- Stewardship of Natural and Cultural Ocean Resources
- Increasing Resilience to Natural Hazards
- Maritime Operations and the Marine Environment
- The Ocean's Role in Climate

- Improving Ecosystem Health
- Enhancing Human Health

In addition, the SOST considered external input in its review of the 20 individual priorities that were designed to advance science in support of the six societal themes. While progress has been made on these areas, the SOST determined that additional progress would provide significant benefits toward societal needs. The updated document adds a set of cross-cutting topics that address prominent issues and priorities that influence every societal theme:

- Global Climate Change
- Social Sciences
- Ocean Literacy
- Ocean Observations and Infrastructure
- Ocean Modeling
- Mechanisms for Collaboration
- International Communication and Cooperation

This new report also updates the societal themes, notably with an increased focus on global change and an emphasis on the Arctic, and highlights both progress and remaining priorities. It also illustrates the direct link between the research priorities and National Ocean Policy priority objectives, noting the foundational relationship of basic research to sound policy and management decisions.

With progress supported by Federal agencies, the research priorities defined in this report form a strong framework for tackling urgent issues that may change through time.

Box 1: Community Input

In updating the ocean research priorities, the SOST worked closely with stakeholders from many sectors. *Science for an Ocean Nation*:

- Incorporates community comments received through:
  - Town hall meetings at national meetings and conferences
  - Dear Colleague Letter in a national publication
- Addresses public comments solicited through the Federal Register on accomplishments and gaps in the original plan and its implementation, and on new needs

- Considers public comments and agency perspectives highlighting sustained priorities and new needs
- Addresses input from:
  – Subcommittee on Integrated Management of Ocean Resources
  – Ocean Research Advisory Panel

Beach near the port of Galilee, Point Judith, Rhode Islands. Credit: Mr. William B. Folsom, National Marine Fisheries Service, National Oceanic and Atmospheric Administration.

The Need for Fundamental Research

Scientific discovery, driven mostly by competitive, peer-reviewed investigations, is a major part of the Nation's research enterprise and is essential for informing the National Ocean Policy. Fundamental research that expands the scientific frontier enhances and deepens our understanding of the ocean and its role in the Earth system. To make progress, it is essential to cultivate and investigate new ideas about the ocean and new approaches for exploring the marine environment that may challenge existing interpretations. In doing so, society should recognize and even encourage risk-taking in supporting the most exciting and promising ideas for making progress in understanding the ocean. Progress requires the continued support of both systematic measurements of the ocean's properties and the freedom to pursue new ideas and technology. Better access to ocean, coastal, and Great Lakes ecosystems made possible by scientific and technological innovation will continue to enable breakthroughs in basic understanding of biology, chemistry, geology, and physics in these ecosystems, as well as the connections among them. Improved access to the ocean, coasts, coastal watersheds, and Great Lakes depends on advances in infrastructure and technology such as advanced sensors, satellites, and unmanned vehicles. The reach of ocean science research can be significantly expanded by leveraging the capabilities of other disciplines, including nanotechnology, robotics, genomics, and other areas to enable new access to and perspectives on the ocean environment. A workforce that is eager to push the limits of scientific knowledge and technological innovation will support the scientific discoveries necessary to address fundamental scientific questions and advance the use of science to support operational activities to meet national needs. By definition, unforeseen breakthroughs and paradigm shifts cannot be planned. Therefore, this document focuses on underscoring—rather than attempting to define or comprehensively enumerate—the fundamental research efforts that provide the foundation for understanding the ocean. It emphasizes the research efforts with particular anticipated societal applications, while invoking the fundamental research that provides the foundation for those applications.

Photo 3. Plankton net deployed to collect near-surface plankton in Maug caldera. Credit: Pacific Ring of Fire 2004 Expedition. National Oceanic and Atmospheric Administration Office of Ocean Exploration; Dr. Bob Embley, NOAA PMEL, Chief Scientist.

OCEAN SCIENCE IN SUPPORT OF NATIONAL INTERESTS

Addressing National Challenges

Our national agenda of social, economic, and environmental goals is focused on ensuring the health, prosperity, safety, and security of all citizens. Progress on this agenda has become more challenging in recent years with the emergence of a series of global issues with national implications and local repercussions. Primary among these are climate change and energy independence, which feed into issues of sustainable economic well-being and national and homeland security. The magnitude, reach, and level of interconnection among these issues require significant and sustained attention. Decisions on how to manage, mitigate, and adapt to the challenges caused by these issues have the potential to alter our economy, our security, our environment, and our daily lives.

In light of these challenges, essential aspects of protecting lives, enhancing livelihoods, and improving quality of life depend on healthy ocean, coastal, coastal watershed, and Great Lakes ecosystems and resources; a secure maritime domain; and prosperous marine-based economic sectors. This reliance is apparent when the direct link between the marine environment and climate, sustainable development, and energy is stated in hard numbers. Fifty-two percent of the Nation's total population lives in our 673 coastal counties. In 2007, these coastal counties contributed $7.9 trillion to the Nation's GDP, more than half of the total. Eight-and-a-half million people reside in the 100-year coastal flood hazard area, 16 percent of whom were below poverty level in 2009.[9] Changing global climate may add to the economic impact of coastal erosion and inundation as sea level rise makes these events even more severe. Over $3 trillion in real estate investments along the Atlantic and Gulf coasts could be better protected through improved building codes and land management practices.[10] Coastal storms account for 71 percent of recent annual disaster losses totaling $7 billion per year, [11] but coastal wetlands provide an estimated $23.2 billion in storm protection services each year.[12] The ocean represents an immense resource for renewable energy that could revolutionize energy production and job creation in the United States. Some estimates indicate that harnessing 1/1,000th of the energy contained in the Gulf Stream's current would supply Florida with 35 percent of its electrical needs.[13] The success or failure of our response to national challenges lies in the information on which we base our decisions. In turn, the quality of that information relies on the science underlying it. Strengthening our understanding of ocean, coastal, coastal watershed, and Great Lakes ecosystems is urgent and essential. Progress on the research priorities outlined in this report will offer powerful support for informed action on our Nation's environmental, economic, and social challenges.

Box 2: Priority Objectives of the National Ocean Policy

1. Ecosystem-based Management
2. Coastal and Marine Spatial Planning
3. Inform Decisions and Improve Understanding
4. Coordinate and Support (Management)
5. Resiliency and Adaptation to Climate Change and Ocean Acidification
6. Regional Ecosystem Protection and Restoration
7. Water Quality and Sustainable Practices on Land
8. Changing Conditions in the Arctic
9. Ocean, Coastal, and Great Lakes Observations, Mapping, and Infrastructure

Beach near the port of Galilee, Point Judith, Rhode Islands. Credit: Mr. William B. Folsom, National Marine Fisheries Service, National Oceanic and Atmospheric Administration.

Informing National Ocean Policy

The National Ocean Policy sets out nine national priority objectives (see Box 2). The first four speak to how we as a country go about using and protecting the ocean, Great Lakes, and U.S. coasts and resources. The remaining five highlight areas where new science, technology, and decision support are especially urgent. Together, they form a general strategy for government and non-governmental stakeholders at all levels to implement the National Ocean Policy. It is a strategy founded on gathering, interpreting, and sharing science-based information to guide actions.

The National Ocean Policy addresses the direct link between stewardship of natural systems and human health, economic well-being, environmental sustainability, national prosperity, adaptation to climate and other environmental change, social justice, foreign policy, and national and homeland security. Its nine priority objectives constitute an initial strategy to address some of the most pressing ocean, coastal, coastal watershed, and Great Lakes challenges, and provide a bridge between the policy on paper and action on the ground.

Underscoring the need for the Nation to continue pursuit of the research priorities highlighted in *Science for an Ocean Nation* is the link between these priorities and the National Ocean Policy. The National Ocean Policy repeatedly emphasizes the need for and importance of science in implementing policy and supporting management decisions.

One of the overarching policy principles is "support of disciplinary and interdisciplinary science, research, monitoring, mapping, modeling, forecasting, exploration and assessment" and one of the priority objectives is "Inform Decisions and Improve Understanding." These specific references illustrate the strong, overarching connection between research and policy. Progress on the ocean research priorities not only advances understanding, but also provides critical support for moving forward on the policy objectives.

As a result of *Charting the Course*, Federal agencies have made particular progress in the four areas identified in that document as Near-term Priorities for initiating rapid progress.

Accomplishments in these areas contributed to the priority objectives outlined in the National Ocean Policy. Details on these efforts and their role in advancing the National Ocean Policy are provided in a following section, "Achieving National Ocean Policy Priorities: Progress through Ocean Research."

Understanding Emerging Topics

Two significant scientific priorities were just emerging as *Charting the Course* was being developed. Their full impact on both ocean science and policy is now being recognized. Research priorities related to these two emerging topics are noted in the societal themes on The Ocean's Role in Climate and Improving Ecosystem Health.

Ocean Acidification

Increasing carbon dioxide (CO_2) in the atmosphere results in increased CO_2 in the ocean, where it lowers pH and changes fundamental ocean chemistry. Organisms that build their shells and skeletons from calcium carbonate are particularly affected. Resulting impacts include increased stress on coral reefs, changes in food chains for commercial fisheries, and disruption of the natural processes that transport and store carbon in the deep ocean. An even more acidic ocean of the future will have a profound impact on marine biodiversity, species viability and distribution, and food webs, but understanding the form those changes will take or their net effect on marine ecosystems is a considerable challenge for adaptation and mitigation.

Science for an Ocean Nation expands the preliminary view of ocean acidification included in Charting the Course and builds on a number of recent events, including inclusion of this topic in the National Ocean Policy and passage of the Federal Ocean Acidification Research and Monitoring Act of 2009. As directed in this legislation, the SOST has established an Interagency Working Group on Ocean Acidification (IWG-OA), which has completed an inventory of agency activities. The IWG-OA is preparing a Strategic Plan for Ocean Acidification Research and Monitoring, which will incorporate science priorities from a growing number of national and international assessments, including one by the National Academy of Sciences, Ocean Acidification: *A National Strategy to Meet the Challenges ofa Changing Ocean*[14] (see Box 3).

Over the past several years, the SOST agencies have accelerated their support for research on ocean acidification in recognition of its growing threat and the lack of basic understanding of biological and chemical responses to acidification of the ocean. Current activities undertaken to advance the understanding of ocean acidification include:

- Research funding announcements with ocean acidification components;
- Increased coastal and ocean monitoring of ocean acidification;

- Research to determine the impacts of ocean acidification on living marine resources, including individual, population, and ecosystem level effects; and research on corals, economically important fishery species, and other ecologically important species;
- Development of advanced remote and *in situ* ocean acidification sensing technologies; and
- Examination of the relationship between ocean acidification and the Clean Water Act.

These activities are identifying areas, species, and ecosystems of concern, as well as potential ecosystem impacts.

This will enable assessment of socioeconomic impacts, development of adaptation and mitigation strategies, and ultimately, improved conservation and management of natural resources.

Box 3: Ocean Acidification

The 2010 National Academy of Sciences report Ocean Acidification: *A National Strategy to Meet the Challenges of a Changing Ocean* noted the overarching need for a "National Program" and highlighted eleven recommendations:

- Create an adequate and adaptive chemical monitoring program.
- Support consistent and clear data acquisition via inter-calibration and standards development.
- Review existing and emergent observing networks to identify and fill critical gaps.
- Plan for long-term sustainability of an integrated ocean acidification observation network.
- Focus on eight unranked priorities:
  - Understand processes affecting acidification in coastal waters.
  - Understand the physiological mechanisms of biological responses.
  - Assess the potential for acclimation and adaptation.
  - Investigate the response of individuals, populations, and communities.
  - Understand ecosystem-level consequences.
  - Investigate the interactive effects of multiple stressors.
  - Understand the implications for biogeochemical cycles.
  - Understand the socioeconomic impacts and inform decisions.

- Identify, engage, and respond to stakeholders to develop a broad strategy for decision support.
- Create a data management office and provide it with adequate resources.
- Establish an Ocean Acidification Information Exchange to provide timely research results, syntheses, and assessments that are of value to managers, policy makers, and the general public.
- Assess facilities and human resource needs.
- Create a detailed implementation plan with community input.
- Create a program office with the resources to ensure successful coordination and integration of all the elements outlined in the Federal Ocean Acidification Research and Monitoring Act of 2009 and this report.

Photo 4. Coral Reef. Credit: National Oceanic and Atmospheric Administration.

Changing Conditions in the Arctic

The National Ocean Policy recognizes that changing conditions in the Arctic create a special focus area for the Nation's research agenda. The Arctic has vast untapped strategic natural resources in some of the most pristine and vulnerable ecosystems on Earth. Though seemingly isolated, this unique region has global influence on both human activities and the natural environment. Changes in the Arctic can have profound implications for global climate change. Fresh meltwater flowing into the ocean can change the ocean circulation system. Melting permafrost releases greenhouse gases to the atmosphere. Diminishing ice cover increases planetary warming as less solar energy is reflected away from the planet. As summer sea ice diminishes, access to the Arctic for tourism, resource extraction, shipping, and other potential activities will increase, enhancing economic prospects and increasing environmental risks. The potential for a significant increase in human activity in the Arctic, coupled with a shifting environment, raises many policy questions. Science is essential to answering such questions but, as noted by the U.S. Arctic Research Commission in its R*eport on Goals and Objectives for Arctic Research 2009–2010,*[15] we are greatly hampered by the lack of observational capability, data, studies, and models in the Arctic (see Box 4).

Box 4: Arctic Research

Supported by multiple Federal agencies with responsibilities in the Arctic, the U.S. Arctic Research Commission calls on the U.S. Arctic Research Program to strengthen efforts in five areas:

- Environmental change of the Arctic, Arctic Ocean, and Bering Sea;
- Arctic human health;
- Civil infrastructure;
- Natural resource assessment and earth science; and
- Indigenous languages, cultures, and identities.

The Commission's Report on Goals and Objectives for Arctic Research 2009–2010 highlights the need for:

- Continued development of an Arctic observing network;
- Further research on urgent environmental change;
- Establishment of baseline oceanographic data;
- Planning for oil spill response in ice-covered waters;
- Addressing impacts and opportunities of increased shipping and energy development; and
- Research infrastructure and human capital.

The U.S. Arctic Research Commission's report, the National Ocean Policy, and ongoing agency activities all support a more focused U.S. Arctic policy. Reviewed and revised in 2009, the U.S Arctic Policy is to meet national security and homeland security needs while protecting the environment, ensuring sustainable development, and practicing precautionary resource management. It highlights the need to "enhance scientific monitoring and research into local, regional, and global environmental issues."[16] Progress being made to address research challenges for Arctic issues includes:

- Ongoing acquisition of data, catalyzed by research activities during the first International Polar Year (2007);
- Continuing development of the Arctic Observing Network;
- Construction of a modern research vessel capable of working at the ice edge; and
- Involving local native stakeholders and their traditional knowledge in environmental assessments and decision making.

A number of agencies are undertaking comprehensive reviews of their Arctic programs and policies, which will feed into strategic actions developed under the National Ocean Policy. Environmental stewardship of the Arctic will require strong collaboration and international partnerships. Research to improve scientific information, assess short- and long-term environmental trends, develop early warning systems, develop systems to respond to pollution incidents, and promote sustainable resource management will be a major component of the strategy.

Photo 5. Brown pelican sitting on top of a sign marking the boundary of the Weeks Bay National Estuarine Research Reserve, Alabama. Credit: National Oceanic and Atmospheric Administration.

SOCIETAL THEMES

Science for an Ocean Nation presents ocean research priorities grouped under six societal themes that encompass significant and compelling issues in key areas of interaction between society and the ocean. Chosen because of their impact across economic, environmental, security, health, and societal concerns, the societal themes present ocean research in terms of societal needs. The research priorities for each theme address the most pressing issues–those that are in need of focused attention to make progress. They include both natural and social sciences and the tools required for carrying out research and translating and disseminating research results. While these research priorities are presented in terms of U.S. needs, it should be noted that many of our national issues are inherently global in nature in that they extend beyond U.S. coasts and waters in their scope and influence. Addressing the full scale of such priorities will necessarily require a global approach in terms of the geographic extent of research and the partnering likely needed to conduct this research and distribute its results.

This report carries forward the research priorities laid out in Charting the Course, identified according to the following criteria:

- Is the proposed research transformational (e.g., will the proposed research enable significant advances for insight and application, even with potentially high risk for its success; would success provide dramatic benefits for the Nation)?
- Does the proposed research impact many societal thematic areas?
- Does the research address high-priority needs of resource managers?
- Would the research provide understanding of high value to the broader scientific community?
- Will the research promote partnerships to expand the Nation's capabilities (e.g., contributions from other partners, including communities outside of ocean science, such as health science; unique timing of activities)?
- Does the research serve to contribute to or enhance the leadership of the United States in ocean science?
- Does the research contribute to a greater understanding of ocean issues at local to global scales?
- Does the research address mandates of governing entities (Federal agencies, state, tribal, and local governments)?

Photo 6. Kemp's Ridley turtle returned to the Gulf of Mexico for release off Cedar Key, Florida, following rehabilitation from oil exposure resulting from the Deepwater Horizon/BP spill. Credit: National Oceanic and Atmospheric Administration.

Theme 1: Stewardship of Natural and Cultural Ocean Resources

Effective stewardship of the tremendous resources provided by the ocean, continental shelves, coastal areas and watersheds, and Great Lakes will ensure that these resources and benefits are available for future generations. These areas provide us with food, water, chemicals, minerals, materials for medical advances, energy, means of transportation, recreation, and tourism. The ocean preserves a record of the world's cultural past in shipwrecks and other archaeological sites, while forming an integral part of America's cultural landscape. The ocean remains a vast, largely unexplored realm with the capacity to provide new pharmaceuticals, industrial products, and renewable energy. At the same time, its resources are subject to many pressures, such as overfishing and bycatch, habitat disturbance and loss, pollution, and competition from invasive species. To unlock the full resource potential of the ocean, continental shelves, coastal areas and watersheds, and Great Lakes, society should improve understanding of these areas and find an appropriate and sustainable balance between resource conservation and use.

Rationale

The U.S. has the largest Exclusive Economic Zone (EEZ) of any nation.[17,18] Our EEZ provides numerous ecosystem services to society.

Commercial fisheries within the EEZ and contiguous inshore waters support about $100 billion in annual sales for the seafood industry.[19] Fishery resources also support the $80 billion recreational fishing industry.[20] Recent advances in aquaculture are increasingly contributing to the seafood supply. While very little of current aquaculture production occurs in the U.S. EEZ there is substantial potential for future development. The total current market value of the oil and gas inventory in our EEZ is approximately $11 trillion, including an estimated $10 trillion in undiscovered, technically recoverable resources.[21] Total gross wind resources off the U.S. coasts are estimated at more than 4,000 gigawatts, or approximately four times the current generating capacity of the U.S. electric grid.[22] Developing wind resources in shallow waters, which are the most likely to be technically and commercially feasible at this time, could provide at least 20 percent of the electricity needs of almost all our coastal States, which use about 80 percent of total U.S. demand.[23] Approximately 200,000 trillion cubic feet of gas hydrate are present on land and the U.S. outer continental shelf (OCS).24, [25] The Great Lakes are the largest source of fresh water in the world.[26] Healthy natural and cultural resources in our ocean and coastal areas are the foundation of a huge coastal tourism and recreation industry that is the fastest growing area of the ocean economy.[27] Marine protected areas, sanctuaries, and designated critical habitats help conserve natural and cultural marine resources as well as attracting tourists. Prehistoric landscapes, shipwrecks, and historic and living waterfronts along the Nation's coasts and Great Lakes all contribute to the national cultural heritage.

The scale and diversity of ocean resources is immense; however, resource use, exploration, and development often compete with other societal needs and values. Overexploitation of many fish stocks and alteration of habitats due to natural and anthropogenic causes have had profound consequences for many ecosystems and fishing communities. Increasing domestic energy production in the EEZ, both renewable and non-renewable, is critical to energy independence but brings with it concerns about environmental impacts. Increasing populations near the coasts and Great Lakes impact freshwater supplies, and can lead to degradation of water quality if wastewater disposal methods are not adequate. Evaluating and addressing the environmental consequences of resource use and extraction, combined with increased understanding of the social and economic value of natural and cultural resources and the factors that influence ecosystem and cultural resilience and health, can help balance the pressures being placed on coastal ecosystems, enable restoration of degraded habitats, and, ultimately, support robust, coordinated ecosystem-based management and governance strategies for sustainable resource use.

Resource management will need to be adaptive in the face of one of the largest challenges facing this generation—a changing global climate. Changing climate conditions are predicted to cause shifts in species distribution, reproductive timing, food webs, sea levels, weather patterns, and ocean chemistry. Thresholds and resilience are unknown for most ecosystems, although scientists know that different environments and species display unique sensitivities to climate variability. The current structure and function of ecosystems and the marine communities they support may change. The immediate impacts of climate change may be most evident in the Polar regions.

In the Arctic, melting sea ice and retreating glaciers are changing the physical environment, opening of the Northwest Passage and increasing access to fossil fuels under the Arctic seabed, some of which may exist in commercial quantities adjacent to our EEZ. Fragile coral reef systems are threatened by coral bleaching and ocean acidification. Coastal wetlands and island communities are threatened by sea-level rise. New scientific research is needed to understand, predict, and adapt to the changes in ocean resources and the consequent effects on our economic and social activities.

The National Ocean Policy identifies ecosystem-based management (EBM) as crucial to stewardship of the Nation's ocean, coastal, coastal watershed, and Great Lakes resources. Compared to conventional resource management, EBM takes a more comprehensive, holistic approach. The key to success of an EBM approach is accurate assessment of the current condition of natural and cultural resources over various spatial and temporal scales, in the context of changing stresses. This ability and knowledge will provide the foundation for understanding the complex relationships between humans and living and non-living resources.

Understanding these connections and the incremental and cumulative impacts of change on environment and human activity can help determine the likely outcomes of various management strategies. Research into issues of resource development, use, extraction, and regional variability will help society prevent major impacts to ecosystems, promote sound development and use of resources, preserve cultural sites and traditions, and support management efforts to restore depleted populations to healthy and sustainable levels. The research should be integrated into a comprehensive, scalable planning approach that balances the principles of EBM with societal uses of ocean resources.

Research Priorities

Progress to Date

Recent progress in scientific understanding of the structure and function of ecosystems is laying a better foundation for sound stewardship. For example, the Comparative Analysis of Marine Ecosystem Organization (CAMEO) program (named a Near-term Priority by Charting the Course) is helping to elucidate key ecosystem processes and the effects of such things as harvesting strategies and climate change on living marine resources. The Census of Marine Life, a ten-year global initiative, has made great strides in assessing and explaining the diversity, distribution, and abundance of marine life. Technological advances, such as the Critter Cam and Hab Cam, radar tracking of seabirds, and more robust marine mammal tags have furthered our understanding of migration patterns, habitat use, and behavior. Capital investments have been made in the National Oceanic and Atmospheric Administration's (NOAA) seagoing fleet including the commissioning and development of new acoustically quiet vessels to improve fish stock assessments and the only Federal vessel devoted exclusively to ocean exploration. Progress has also been made in our ability to assess energy and mineral resources and in the development of environmentally safe, renewable energy technologies. These scientific and technological advances have been bolstered by recently enacted legislation that supports stewardship of marine resources.

Photo 7. Divers examine wreckage of 1800's Nantucket whale ship, Two Brothers, off French Frigate Shoals, approximately six hundred miles northwest of Honolulu, Hawaii. Credit: National Oceanic and Atmospheric Administration.

Updated Priorities

There has been recent progress in the stewardship of natural and cultural resources, but there is still need for continued efforts.

Research Priority 1: Understand the Status and Trends of Resource Abundance and Distribution through More Accurate, Timely, and Synoptic Assessments

- Assessing the coupled impacts of resource use and extraction (e.g., fisheries, ocean mining, tourism, energy) and systemic change, we should first measure the abundance and distribution of living and non-living resources in the open ocean, continental shelves, coastal areas and watersheds, and Great Lakes. To make these measurements, we need the tools and capabilities that will allow us to:
- Assess the status and health of fish stocks and protected resources;
- Monitor living resources (at multiple trophic levels) using both fishery-independent and fishery-dependent data collected nearly continuously over wide swaths of ocean at appropriate levels of species resolution;
- Assess the spatial and temporal variability (both natural and use-induced) of resources (e.g., biota biodiversity, energy, minerals, pharmaceuticals), including in deep-water settings; and
- Provide sustained monitoring and mapping of natural and cultural resources to understand and better identify physical, biological, and social thresholds and shifts.

Developing and implementing these tools and capabilities in a systematic, proactive manner will not only improve the protection of resources, but may be more cost-effective compared to reactive and curative responses after disturbances or losses have already occurred. For example, consider the tremendous expense and economic dislocation associated with rebuilding depleted fish populations.

Improved assessments will not only increase our capability to proactively reduce these costs, but also increase understanding of and our ability to minimize the secondary effects of harvesting, such as fisheries-induced changes to genetic diversity and population structure, loss of biodiversity, and bycatch of non-target species. Climate change necessitates that systematic observation of the status and trends of resources begin quickly to allow for the determination of system resilience and response as well as thresholds.

This approach will enable adaptive long-term management of natural and cultural resources, using empirical evidence to design and modify management and marine planning strategies (e.g., individual harvest quotas, special use areas, marine protected areas).

Research Priority 2: Understand Interspecies and Habitat–Species Relationships to Support Forecasting of Resource Stability and Sustainability

Understanding the individual, incremental, and cumulative impacts of an ever-widening array of human activities (e.g., transportation, recreation, land use, fishing, energy exploration and development, increased food production from aquaculture) on water quality, habitat, and the ocean's ability to support marine life and biodiversity is a necessary foundation for ecosystem-based management.

Often the data necessary for these assessments are difficult to obtain because of practical limitations on controlled experimentation and replication in natural systems.

Investing in observations, habitat characterization and mapping, process studies, and advanced modeling will expand current understanding of impacts at appropriate temporal and spatial scales, and help identify crucial data and gaps in process-understanding so that appropriate resource-management techniques can be developed and implemented.

Priorities to enable improved natural-resource forecasting include:

- Collection of necessary data (observational and experimental) to support robust models;
- Development and validation of ecosystem and species interaction models at appropriate scales that incorporate feedback mechanisms among trophic levels;
- Assessments of regional and local models of environmental changes (e.g., responses to annual, decadal, and longer-term climate drivers; rapid regime shifts; hydrodynamic circulation; watershed discharge) to help determine how these changes impact resources; and
- Development of approaches and scenarios to understand and integrate the specific and cumulative impacts of various natural resource policies on living resources and human communities; e.g., incorporating improved assessments of impacts on ecosystems (see Theme 5, Improving Ecosystem Health).

Research Priority 3: Understand Human-Use Patterns That May Influence Resource Stability and Sustainability

Determining the value of natural and cultural resources and evaluating the effects of alternative management scenarios require consideration of economic (market and non-market), sociological, and cultural knowledge, including potential competing uses. These assessments, coupled to the scientific evidence developed through other research priorities, will also support evaluations of the socioeconomic tradeoffs inherent in management efforts. Human-use data can be acquired through surveys and analysis of economic and sociological factors associated with a broad range of resource activities in the open ocean, continental shelves, coastal areas and watersheds, and Great Lakes. Socioeconomic data (e.g., resource cost and earning, demographics, income) that is geocoded (linked to geographical locations) can be combined with new methods of assessing non-market valuation (see Theme 5, Improving Ecosystem Health) to integrate socioeconomic and natural processes in models to determine their interactions. Social, cultural, and economic research, combined with resource-use assessment and modeling, will support effective evaluation of how various management options will impact resource use and sustain-ability. As management options can be more accurately evaluated, the balance points between resource use and conservation can be better determined. Improving outcomes of resource management through the effective application of marine planning is unequivocally linked to both public education and the education of the next generation of scientists and planners. The effectiveness of current educational approaches used to promote stewardship and achieve sustainable use of resources should be evaluated.

Research Priority 4: Conduct Applied Research Using Advanced Understanding to Enhance the Benefits of Various Natural Resources from the Ocean, Coasts, Coastal Watersheds, and Great Lakes

Key technological advances are required to facilitate long-term and effective resource development and management; to make more responsible use of available renewable and non-renewable resources, such as energy, minerals (including sand), and pharmaceuticals; and to support alternatives to resource extraction (e.g., aquaculture, renewable energy), along with efforts to improve the efficacy of marine operations while simultaneously limiting their impacts.

Areas of focus include, but are not limited to: Development of sustainable approaches to aquaculture that consider the implications for surrounding ecosystems, wild genetic resources, and impacts on coastal economies;

- Advancement of new sustainable energy technologies from the ocean, including efficient methods of power generation (e.g., wind and tidal power) and scientific information to fully consider impacts on marine life and marine processes;
- Development of a new generation of military subsurface detection systems that are less deleterious to species possessing sensitivities to acoustic emissions in certain sound ranges;
- Development of bycatch-reduction technologies for fisheries and protected resources (e.g., seabirds);
- Development of innovative economic and social research techniques and methodologies applicable to ocean and coastal industries and communities; and
- Use of advances in exploration technologies (e.g., geospatial, acoustic, bio-optical) to map and characterize the U.S. EEZ and continental shelf, and establish the outer limits of the U.S. continental shelf where it extends beyond the limits of the EEZ and assess important ecological and economic resources found there.

Tools to be Emphasized

Continued advancement in technologies and procedures would enable rapid, efficient, and synoptic assessment of ocean resources (including inventories and alterations). Right-sizing and improving the capabilities of the U.S. domestic fleet, including opportunities for covering all regions, would improve our ability to provide important information on the status of managed populations and the ecosystem effects of human activities. Additionally, improved *in situ* and remote-sensing tools would obtain necessary biological (e.g., species composition, abundance, and movements) and physical (e.g., current direction, wave height) parameters; track living marine resources throughout their ranges; detect changes in ocean character and biodiversity; and survey deep waters, particularly as energy exploration moves to the edges of the OCS and into previously inaccessible regions of the Arctic.

Advances in unmanned aerial and autonomous underwater vehicles, as well as the use of marine animals as living platforms for data collection, will aid in collecting synoptic resource information. In addition to monitoring and assessment with existing technologies, new and improved technologies and protocols will expand the capabilities of fixed and mobile observing platforms to include biological and chemical sensors and advanced acoustics and mapping capabilities.

Improved ecosystem models would improve understanding of complex ecosystem dynamics and forecast the effects on ecosystems of resource use, exploration, and development. Improving these capabilities will further knowledge of the effects of climate change, habitat alteration, resource extraction, and the interactions of ecosystem components. Refining models will enable the prediction of emergent properties such as resilience and thresholds, as well as possible future states of ecosystems given current trajectories. Decision-support tools based on scientific research will assist in evaluating the impacts of human actions and the potential outcomes of different management strategies. There is a pressing need for "end-to-end" applications of science that translate sound scientific knowledge into a format that can be used as a basis for management decisions including visualization and valuation capabilities. Such applications will allow resource managers and stakeholders to make informed decisions in marine planning and in allocating space for competing uses. Partnering with various marine industries to undertake competitive and collaborative research is also important, particularly given the expense of marine research.

Improved information technology, data-support infrastructure, and data management policies are also a priority. For example, biological resource databases are generally not well-integrated with physical and human-use data. Comprehensive environmental databases with appropriate spatial and temporal resolution that integrate information from local, regional, national, and global sources are key to understanding ecosystem interactions and supporting effective and adaptive resource management. Geographic information systems with four-dimensional (latitude, longitude, depth, and time) predictive capability can integrate diverse data sets (e.g., physical, biological, economic, social, cultural) for spatial analysis, static and dynamic modeling, and improved scientific collaboration.

Development of integrated ecosystem assessments, a priority detailed in the National Ocean Policy, will, in turn, expand understanding of interrelationships among the physical environment, ecosystems, and human activities and allow development of end-to-end applications for applying that understanding to decision making.

Implementation of these new approaches requires more technical and scientific personnel skilled in diverse disciplines including natural sciences, social sciences, and information technology for the operation, maintenance, and interpretation of information. Investment in training and maintaining this next-generation workforce is a key priority.

Regional Applicability

Open ocean, continental shelves, coastal areas and watersheds, and Great Lakes regions in the United States support a diverse set of living and non-living resources. Societal views of these resources often vary regionally, necessitating regionally different approaches to management. For example, the Gulf Coast is a major supplier of oil and natural gas, and marine planning there will have to balance energy production and exploration, habitat for commercially and ecologically important fisheries, sustainable development, and increasing resilience to natural disasters. The Pacific Islands and Caribbean are home to coral reefs threatened by rising water temperatures, overfishing, and ocean acidification. The Arctic region contains a variety of largely untapped resources in a changing physical environment, which is becoming more accessible every day. Additionally, native Arctic communities survive largely on subsistence harvesting of marine species. Differences between regions are not restricted to ecosystems but extend to management structures and cultures as well. Certain agencies may lead in one region, but not in others. In spite of this, the overarching need for comprehensive cooperation and collaboration is consistent throughout the Nation. All invested stakeholder groups need to be incorporated into the decision-making framework. The Puget Sound Partnership in Washington State is an excellent example of Federal and State governments, scientists, local tribes, and other citizens working together to improve the health of a regional ecosystem.

Theme 2: Increasing Resilience to Natural Hazards and Environmental Disasters

Impacts of natural hazards and environmental disasters have enormous economic, environmental, social, public health, and homeland and international security consequences. With over 50 percent of the U.S. population living in coastal counties and that number rising each year,[28] the United States should be adequately prepared for coastal disasters. Even as the Nation continues to direct significant Federal funding toward science and technology to assist in reducing the impacts of natural hazards, we face enormous losses each year from ocean and coastal hazards, including:

- Severe storms, hurricanes, tornadoes, and associated coastal and offshore wind, wave, and current damage;

- Coastal inundation and flooding from storms and regional meteorological events;
- Earthquakes, landslides, volcanic eruptions, and resulting tsunamis; and
- Catastrophic pollution events, such as oil spills.

Not to be forgotten are the substantial costs of ecological degradation and public health threats associated with secondary impacts, notably the mobilization of contaminants, disease organisms, and debris materials following hazard events.

Rationale

Every natural disaster provides an opportunity to examine and improve capacity to forecast, assess, and mitigate, and respond to hazard events. Since the completion of *Charting the Course*, several natural disasters have provided devastating and deadly reminders of the continuing need for improved tools and technologies to prepare for, mitigate, and respond to hazard events. In 2008, Hurricanes Ike and Gustav struck the highly developed, highly vulnerable U.S. Gulf Coast. In 2009, the Samoan earthquake generated a tsunami that devastated a community and the ocean and coastal ecosystems on which that community relied. In 2010, earthquakes in Haiti and offshore of Chile took thousands of lives and forever changed communities, including those with marginal capacity to effectively respond and rebuild. That same year, the Deepwater BP oil spill in the United States caused one of the most severe environmental disasters that has ever occurred in this country (see inset). The substantial cost of these events, in lives lost and communities devastated, coupled with the long-term over-development of coastal communities, highlight the need for continued efforts to assess risk and vulnerability and to use that information to prioritize, develop, and implement effective and tailored community awareness programs and warning systems. As the pace of economic, ecological, and social recovery efforts subsequent to the above events shows, the resilience of impacted communities depends on the resilience of critical social and ecological services. Scientific assessments of hazards should include impacts beyond the immediate loss of life and infrastructure in order to fully grasp both the vulnerability and the resilience of at-risk communities and ecosystems.

Effective response to natural hazards is impeded by limitations in our ability to comprehensively assess hazard potential, vulnerability, and risk. Human and natural processes (e.g., land development, sea-level rise) constantly alter system vulnerability and risks to coastal communities. Comprehensive risk and vulnerability assessments should now consider climate change impacts including sea-level rise and resulting inundation, changes in coastal storm pattern and severity, sea ice loss, and degradation of coral reefs due to ocean acidification. This direct connection between a changing natural environment and the safety of coastal communities, infrastructure, operations, and ecosystems—and what this means to economic, social, health, and environmental security—is increasingly recognized among citizens and decision makers. These elements comprise the concept of *coastal resilience* and focus our attention on its importance. To reduce the social and economic impact of natural hazards, the public and decision makers should be educated about the risks, associated impacts, opportunities for mitigation, and the costs of responding to and not preparing for natural hazards. A sound scientific and technological basis for making decisions, such as that created through marine planning, should include improved models of hazard impacts, more accurate and timely forecasts, and thorough assessments of future hazard potential, vulnerability, and risk. In-depth scientific evaluation and classification of current social institutions will allow decision makers to build on successful efforts while identifying critical gaps in community resilience.

Engineering, insurance, and preparedness solutions to coastal hazard mitigation begin with consistent, state-of-the art forecasting and observation. Hazard assessments should be comprehensive, reflecting multiple hazards on very different timescales and their impacts on diverse economic, social, and ecosystem sectors. Assessments of physical processes should be coupled with improved understanding of the role of human behavior and social systems, especially in view of existing institutions and infrastructure. This calls for information on ecosystem health (e.g., response to resource and land use), impacts of climate change (e.g., sea-level rise, storm intensification and/or concentration), and direct and indirect risks to public health, including from secondary dangers and loss of public sanitation infrastructure. The infrastructure and activities necessary for a range of marine operations are at risk from hazard events. Maintaining or reestablishing marine operations, including those that are seaborne or seafloor-based, during the course of and after a hazard event is critical to every aspect of the Nation's security. Improved hazard understanding, effective research translation, communication, and education are integral for developing a "risk-wise" population, supporting cost-effective strategies designed to increase resistance, enhancing resilience to hazards, and promoting avoidance.

The National Ocean Policy sets forth nine priority objectives—covering topics ranging from ecosystem-based management to changing conditions in the Arctic—as a cornerstone for initiating implementation of the policy. Increasing resilience to natural hazards directly intersects all of these nine objectives.

Photo 8. Deep Ocean Assessment of Tsunami (DART) Easy to Deploy (ETD) monitoring systems are positioned throughout the ocean for real-time tsunami forecasting. Credit: Pacific Marine Environmental Laboratory, National Oceanic and Atmospheric Administration.

Research Priorities

Progress to Date

Since the release of *Charting the Course*, advances in scientific understanding, technology, and tool development have helped the Nation make great strides toward increasing resilience to natural hazards. These advances include probabilistic tsunami assessments for certain locales, an improved tsunami warning system observations, hurricane vulnerability assessments, improved inundation framework (observations and models) at local and regional scales, improved understanding of tsunami and earthquake hazards in the Caribbean and along the U.S. east coast, and improved unified earthquake probability and shaking hazard calculations for California. Through the *Charting the Course* Near-term Priority Forecasting the Response

of Coastal Ecosystems to Persistent Forcing and Extreme Events, progress has been made both in research and decision-support tools. A partnership among Federal agencies and private entities is supporting research to address climate and natural hazard resilience. Another Federal partnership is developing visualization tools that support decisions regarding inundation of the Delaware and Alabama/Mississippi coasts. Yet another partnership is developing tools to support sustainable development and site selection for public housing. This Near-term Priority is also serving as a focal point for improving and leveraging relevant sea-level rise and inundation efforts across the U.S. Gulf region. Progress has also been made through the "Great Shakeout," an annual hazard-scenario event that engages Federal, state, and local agencies and millions of citizens. This exercise highlights the complex interactions among hazard events, resulting secondary hazards (e.g., fires, mobilization of contaminants), critical infrastructure and services, and public and private response; and helps to identify improvements and continuing needs in addressing hazard events. Legislative actions since *Charting the Course* was released also look to provide avenues for increasing resilience to natural hazards. The Ocean and Coastal Mapping Integration Act calls for a coordinated ocean, coastal, and Great Lakes mapping plan to assist with decision making and inform the siting of research and other platforms. This will provide valuable support to marine planning efforts. The Integrated Coastal and Ocean Observation System Act calls for the establishment of an integrated national observing system to provide plentiful and diverse information to improve assessments of hazard risk, vulnerability, and resilience.

Updated Priorities

Despite these advances, much remains to be done in achieving timely, reliable assessments and forecasts of risk and vulnerability, and increased community awareness and ability to prepare for and rebound from natural hazards, including protecting and rehabilitating ecosystems. Research priorities should consider the temporal scales of hazards, their spatial distribution relative to population, and overall expected losses.

Recent hurricane and extreme storm losses have been somewhat offset by the insurance industry, whereas strong shaking and tsunami losses would be borne almost entirely by society as a whole. Loss of life and property would likely be greatest from a large coastal earthquake, but the repeat intervals of such events can be on the order of hundreds of years.

Research Priority 1: Understand How Hazard Events Initiate and Evolve, and Apply That Understanding to Improve Forecasts of Hazard Events

Quantifying future hazard potential requires understanding the generation of hazards (e.g., storms, submarine and coastal landslides, tsunamis, flooding) and past hazard occurrences. Improved hazard forecasts, particularly of storm formation, track, intensity, and associated waves, surge, and flooding, will support more effective mitigation planning for and response to hazard events. Priorities for acquiring the necessary basic understanding and applying it to improved forecasting include:

- Identifying, mapping, and monitoring active submarine faults to help assess potential offshore sources of quakes;
- Mapping and coring submarine landslides to determine their volumes, ages, and geotechnical parameters, and to evaluate their potential to cause tsunamis;
- Detailed mapping of near-shore bathymetry, essential for creating models of tsunami run-up;
- Analysis of the geological record of tsunami and storm deposits, an important tool for assessing the recurrence interval of these hazards;
- Creation of kinetic and dynamic models of the interactions between tectonic blocks and faults to provide a framework in which earthquake, tsunami, and landslide hazards may be evaluated; and
- New methods of calculating the probability of earthquake sources that better reflect current knowledge of rupture processes and that incorporate space-based geodesy (earth measurement), as well as quantitative analysis of high-resolution bathymetry and coastal topography to more accurately quantify the odds of strong shaking and tsunami inundation.

Research in this area should focus on process studies and the development and validation of hazard models, including hazard generation (e.g., tsunami sources, seafloor stability, storm formation) and evolution (e.g., tsunami propagation, storms, inundation). The interagency Hurricane Forecast Improvement Plan seeks to provide improved forecast skill. Probabilistic models and assessments of hazard potential and vulnerability should include the effects of land subsidence and future climate change (e.g., changes in storm intensity or frequency, sea-level rise, landscape change) and better characterization of uncertainty in their frequency and severity.

Research Priority 2: Understand the Response of Coastal and Marine Systems to Natural Hazards and Apply That Understanding to Assessments of Future Vulnerability to Natural Hazards, Including the Increased Vulnerability Caused by Climate Change

Coastal ground is often formed from recently deposited, poorly consolidated, highly saturated sediment. It is therefore especially vulnerable to strong shaking, liquefaction, landslides, debris flows, wave forcing, and erosion. As coastal populations grow, it is necessary that better hazard and risk modeling specific to coastal settings be conducted. Natural hazards impact infrastructure directly through alterations to the underlying landscape and through secondary processes (e.g., slope failures, shoreline change, inlet formation, coastal erosion, sediment transport, flooding). Natural hazards also have significant impacts on coastal features, such as shorelines, as well as cascading and nonlinear impacts throughout ecosystems, during and after the hazard event. Natural systems, such as wetlands, reefs, and mangrove forests, play a significant role in coastal resilience and surge attenuation. Understanding the capacity of these systems to mitigate the effects of natural hazards, how altering the system (i.e., through physical destruction and sediment diversion) affects this capacity, and how the dynamics and processes of altered ecosystems will improve our ability to facilitate recovery of ecosystem functions and to undertake appropriate restoration efforts. Priorities include:

- Understanding and modeling landscape change associated with coastal hazards, including impacts across the adjacent watershed;
- Understanding how the hydrodynamics of extreme flooding events (e.g., tsunamis) are affected by vegetation;
- Understanding resilience of both natural structure and built infrastructure; and
- Assessing vulnerability of coastal communities, public health, infrastructure, marine operations, and ecosystems to primary and secondary hazard processes.

Models of response and vulnerability should be relevant to future conditions, reflecting landscape, ecosystem, and community responses to climate change and the consequences of evolving development and resource use. Models based on observations of pre-event conditions and post-event impacts are required to forecast the magnitude and nature of impacts, the changing vulnerability of altered landscapes (e.g., the influence of land use and restoration on a system's ability to resist or adapt to hazards), and long-term impacts on ecosystem functions and health.

Research Priority 3: Develop Multi-Hazard Risk Assessments and Support Development of Models, Policies, and Strategies for Hazard Mitigation

Integrating observations, data, and models will improve understanding of hazard potential and vulnerability, and improve risk assessments for coastal communities, as well as public health and safety, infrastructure, marine operations, and ecosystems. Research efforts should identify vulnerable ecosystem functions and infrastructure components, determine the potential for cascading component failure, and assess the efficacy of natural systems (e.g., barrier islands, coastal wetlands) and engineered systems (e.g., hurricane barriers, levees) in hazard mitigation. Cascading failure initiated by earthquakes can involve severe coastal ground shaking, liquefaction, submarine landslides, failures of coastal cliffs, and the generation of tsunamis and tsunami-like waves. Similarly, sudden magmatic injection events and volcanic eruptions can be accompanied by significant seismic activity and the destabilization of coastal regions, which can act as secondary triggers for tsunamis. Probabilistic frameworks for multi-hazard assessments need to be better developed to encompass these multiple component failures and be integrated with improved understanding of system responses to hazards and incorporated into comprehensive risk assessments. The degree to which multiple components of hazards are linked physically (rather than simply overlapping spatially) should be formalized and expressed in usable terms. Such risk assessments will support the identification of effective and affordable systems, materials, and technologies for hazard-resilient and resistant communities, infrastructure, and ecosystems, as well as supporting development and management approaches that minimize hazard impacts.

Models of resilience and resistance to hazards should also address social and economic drivers that influence pre-hazard activities (e.g., land use, shoreline development) and responses to hazards (e.g., warning responses, community development, economic restructuring). Integration of models and risk assessments into decision-support tools will inform short- and long-term public and policy responses to hazards, minimizing the economic, social, and environmental costs of natural hazards. Integrating natural and social sciences will ensure that hazard products and policies address public perceptions of risk, communicate the consequences of community and individual responses to immediate and long-term hazards, and encourage appropriate public response to hazard assessments and warnings ("risk-wise" behavior).

Photo 9. Tsunami damage in Santa Cruz Harbor, California. Credit: National Ocean Service, National Oceanic and Atmospheric Administration.

Research is required to develop the methods and tools necessary to create consistent regional-scale integrated assessments that effectively translate scientific understanding, including uncertainty, to management decision-making. This will ensure that risk and vulnerability assessments reflect current conditions and anticipate the impacts of climate change and development decisions. Efforts are also needed to make site-specific assessments of the risk to, and vulnerability of, critical infrastructure, adjacent communities, and natural resources. This will advance an integrated multi-hazard perspective including both global and local processes spanning extreme (e.g., storms, earthquakes, tsunamis, slope failures) to persistent (e.g., erosion, sea-level rise) hazards acting on the seabed, shoreline, and coast.

Tools to Be Emphasized

Reducing hazard impacts and increasing hazard resilience will require continued expansion of existing observation systems, data-delivery systems, and modeling capabilities at regional, national, and global scales. Sustained and integrated marine and coastal observations, including technological improvements in remote and *in situ* sensing of multiple time-sensitive parameters (oceanographic, geophysical, hydrological, chemical, biological, geographic, and anthropogenic) and comprehensive geospatial characterization combined with sustained and reactive-deployment modes of operation are required to acquire baseline (pre-impact) and post-hazard data.

These deployments should be capable of assessing diverse hazard processes and impacts, such as coastal and inland inundation and water quality. These deployments would support the development of improved models of (for example) storm evolution and coastal response. The capability to provide up-to-date characterization of coastal vulnerability and pre-storm conditions (including topography and bathymetry, land use and cover, and riverine and coastal water levels) should be maintained. Multi-hazard assessments will require an information infrastructure that supports a geographic information system (GIS) that includes all hazards, and has defined standards for information integration and application. Advanced data-integration and computational resources and the establishment of community computational standards are required for data-assimilation systems, coupled land–ocean–atmosphere models, and data management and delivery systems.

Ocean literacy is perhaps the most critical aspect of increasing resilience to natural hazards. Building community awareness through appropriately-packaged information products is key. Increasing resilience hinges on understanding community needs and enabling community members to act. Social science data on topics such as economic reliance on resources and habitats, existing social structures, and avenues to ensure robust communication are a necessity.

The design and implementation of robust, reliable, and widely available notification systems will help enable more effective communication of warnings. This will require an end-to-end process for moving from data to tools. Data collection should include social, economic, and educational information to understand at-risk communities, and traditional and historic knowledge from the community to augment scientific observations and provide information on frequency of events. Deterministic and probabilistic forecast models, time-dependent hazard models (including models on timescales relevant to climate change), multi-hazard models, and models of cascading effects are then needed to forecast hazard events and their impacts. There is an essential requirement for a workforce capable of obtaining and blending the physical, mathematical, and social data into decision-making tools, including regionally-appropriate assessments of risk for decision making by the public (coastal and resource managers) and private sectors (insurance, reinsurance, and marine operations), and forecast tools that translate warnings into effective responses at local, regional, national, and international levels.

Research communities that address hazards currently focus on single hazards, single elements of the hazard cycle (e.g., generation, impact, policy), or single research disciplines (e.g., physical, ecological, social, and engineering sciences). These diverse communities should be coordinated and integrated to allow the broadest consideration of costs and effectiveness (i.e., on economic, sociological, and ecological levels) to inform risk reduction strategies. This integrated approach will lead to more effective translation of research results and development of tools for informed and comprehensive decision making by public and private sectors.

Given the global prevalence of hazard events, international cooperation is key to understanding risk, vulnerability, and capacity for resilience, and to improving warning systems and coastal communities' ability to appropriately respond to warnings. International cooperation is also critical in mobilizing and funding humanitarian aid following hazard events. The need for cooperation extends to mechanisms for interaction across Federal agencies and among different levels of government. A broad and comprehensive hazards research enterprise depends on partnerships across the natural and social sciences and with policy-makers and representatives of communities at risk. The SOST has engaged the National Science and Technology Council (NSTC) Subcommittee for Disaster Reduction (SDR) to assist in developing a joint focus on coastal inundation. This partnership is critical to ensuring a coordinated and effective Federal endeavor. The SDR has also produced *Grand Challenges for Disaster Reduction*,[29] which discusses coastal inundation as well as hurricanes and tsunamis.

Regional Applicability

Hazard potential, vulnerability, and capacity for resilience vary across regions depending on the natural and built environments, the social infrastructure of at-risk communities, and the extent to which at-risk communities rely on the natural environment. Each region has unique ecological, social, economic, and educational factors to consider, resulting in the need for diverse approaches across regions to address risk and mitigate impacts. Among the primary considerations in developing response and mitigation strategies is the way in which climate change influences hazard events. Understanding ocean–climate interactions within and across regional scales directly supports our ability to better understand and forecast hazard risk and vulnerability. We require national protocols with regional implementation and prioritization.

Since *Charting the Course* was released, existing regional alliances have matured and others have been established. This regional structure should be mobilized in determining and addressing regional needs.

Box 5: Deepwater Horizon Oil Spill

On April 20, 2010, the Deepwater Horizon drill rig exploded, burned, and eventually sank to the bottom of the Gulf of Mexico, resulting in the loss of 11 lives.

Damage to the well it was drilling, located in over 1,500 meters of water, resulted in the largest United States oil spill in history, liberating over 200 million gallons of light Louisiana crude oil. The massive response, clean up, natural resource damage assessment, and restoration process highlight many of the societal themes and points of emphasis in *Science for an Ocean Nation*:

- The limited ocean observing capability in the Gulf of Mexico at the time of the explosion required the shifting of substantial observing resources (ships, aircraft, gliders, and autonomous underwater vehicles) from other high-priority tasks within the Gulf and around the Nation.

  These other high-priority science missions thus went unaccomplished. The spill response highlights the importance of a robust deployed ocean observing capability (including all the assets previously noted).
- Ocean trajectory models for surface and sub-surface oil tracking were useful in short-term response efforts, but less so for medium- to longer-term spill trajectories and impacts. More realistic models of sub-surface trajectories would result from understanding of the size and buoyancy of oil particles, their rates of microbial degradation, and ocean dynamics including mixing.
- Impacts on living marine resources include direct effects on abundance and contamination, as well as indirect impacts on reproduction, fate of eggs and larvae, and sub-lethal damage due to the oil and the use of dispersants.

  New approaches are needed for sampling and rapidly assessing living marine resource populations and impacts of a wide variety of anthropogenic stressors, including synoptic surveys, use of animal-borne sensors, and new sensing technologies.

- The loss of ecosystem services (fishing, tourism, and recreation) resulted in large and consequential impacts on communities. Application of economic, cultural, and behavioral sciences is appropriate to gauge the degree of resilience of these communities to such events, and to underpin compensation of the public for damages under the Natural Resources Damage Assessment process.
- The death of 11 offshore oil industry workers is a reminder of the inherent dangers involved in many fields of marine work. Improved safety for maritime operations is a critical outcome of the oil spill investigations.
- The extensive media coverage of the oil spill, including descriptions of ecosystems, oceanographic processes, and the role of oil production in the local community and the Nation, presented an opportunity to educate the public not only regarding ocean processes but on the trade-offs in decisions on resource use and management.
- The oil spill is one of a number of significant, simultaneous stressors affecting the Gulf of Mexico Large Marine Ecosystem. Restoration of the ecosystem and understanding the interplay of nutrient enrichment, habitat loss, effects of hypoxic conditions, effects on fisheries, and the impacts of oil and gas development require a quantitative framework. Ecosystem modeling leading to quantitative decision-support tools could have important benefits in prioritizing recovery investments.

Photo 10. Deepwater Horizon site. Credit: National Oceanic and Atmospheric Administration.

Photo 11. USS Antietam (CG 54) departs San Francisco Bay. Credit: U.S. Navy photo by Lt.j.g. Pete Lee/ Released.

Theme 3: Maritime Operations and the Marine Environment

Priorities under this theme focus on understanding the interactions between maritime operations and the marine environment, and using that understanding to predict changing conditions in the maritime environment. Maritime operations are essential services that support vital elements of economic prosperity, national security, and homeland defense as well as the global economy.[30] Maritime operations require freedom of navigation in the global ocean; ability to conduct mapping and charting activities; use of ocean resources; and collection, modeling, and presentation of environmental data. Maritime operations also encompass more narrow issues such as ballast water discharge and assessing ice coverage. Forward-thinking, innovative research, coupled with technological advances, will permit marine operations to address environmental concerns and to meet challenging requirements for increased levels of transportation and commerce in the maritime domain. These must be accomplished while balancing sustainable use and protection of the environment as it undergoes global change.

Rationale

Maritime operations comprise a broad set of activities that occur in the ocean, coasts, coastal waterways, and Great Lakes. These include merchant ship traffic, military operations, fishing, aquaculture, oil and gas drilling, pipeline construction, public transportation, offshore renewable energy projects, seabed mineral extraction, and placement of underwater communication cables. Each of these operations can create adverse interaction with the environment if not performed safely and with consideration of environmental sensitivities. Reliable weather and sea-state forecasts and up-to-date maps , charts, and navigation capabilities to support safe and effective operations and to foster improved maritime domain awareness, will preserve public health, economic security, and the marine environment.

The predicted significant impacts of climate change on the marine environment (e.g., frequent and severe coastal storms, an ice-diminished Arctic, and accelerated sea-level rise) point out the importance of improved observation and modeling of environmental factors and their potential effects on safe and effective marine operations. While we can anticipate some of the likely impacts that these changes will have on maritime operations (e.g., impeded shoreline transit, possible reduction of fish catch due to ocean acidification), more research is needed to determine their long-term impacts.

Particularly in the Arctic, where changing conditions have already diminished ice, a greater understanding of current and potential changes in navigable passages will allow us to predict new shipping patterns and assess their impacts on the natural environment. Research is needed to improve forecasts necessary for safe maritime operations, including atmospheric and ocean conditions, such as sea height, wind, ice conditions, albedo, and severe storm forecasts. This will require more refined climate monitoring, data collection, modeling, and presentation to the mariner.

Maritime operations are also changing with the emergence of new uses and expansion of existing ones. The majority (by tonnage) of the Nation's commerce travels through our maritime ports. Projections state that trade will increase, ships will become larger, and populations along coasts will continue to grow. It is anticipated that offshore oil and natural gas development, with appropriate regulatory, technical, and safety improvements, will continue in the Gulf of Mexico. At the same time, technology is improving to allow new uses of U.S. waters, including ocean renewable energy projects and offshore aquaculture, as the Nation tries to engage in more sustainable energy and fishing practices. Each of these will have impacts on the marine environment and the Nation's ability to conduct marine operations.

To effectively use the results of the research outlined below requires improved communication and collaboration among the diverse stakeholders involved in marine operations, including industry; local, tribal, state, and

Federal governments; and researchers. Avenues for this coordination are suggested in the National Ocean Policy.

Photo 12. Port of Seattle. Credit: National Oceanic and Atmospheric Administration.

The increases in data collection, modeling, and forecasting that will facilitate maritime operations are also necessary components of several priority objectives in the National Ocean Policy: marine planning; informed decisions and improved understanding; and ocean, coastal, coastal watershed, and Great Lakes observations and infrastructure.

Both the Integrated Coastal and Ocean Observation System Act (ICOOSA) and the National Ocean Policy emphasize the need for better ocean observation and infrastructure to guide science-based decision-making. Many of the nine priority objectives in the National Ocean Policy (e.g., marine planning, resilience and adaptation to climate change and ocean acidification, and changing conditions in the Arctic) will require the use of *in situ* observations and modeling.

Increased understanding of the effects of the environment on maritime operations, and vice versa, will facilitate finding a balance between operations and environmental protection. Additionally, more effective collection, processing, and modeling of environmental data will ensure safer and more secure operations.

Research Priorities

Progress to Date

Since the original publication of *Charting the Course*, significant progress has been made toward the research and Near-term Priorities related to marine operations. Many of the activities are being undertaken by the Committee on the Marine Transportation System (CMTS), a partnership of Federal agencies that coordinates with the National Ocean Council and addresses issues pertaining to the marine transportation system (MTS). The CMTS has been addressing research and development and advances in navigation technologies through two integrated action teams. The first of these teams, the Research and Development Integrated Action Team (IAT), co-led by the U.S. Army Corps of Engineers and NOAA, is developing an inventory of Federal MTS research and development activities as a step toward completing a report on Federal priorities for MTS research and development to complement *Charting the Course*. The second team, the Navigation Technology Coordination and Integration IAT, led by NOAA, is pursuing applications for electronic navigation. Under the Energy Policy Act of 2005, the Minerals Management Service (MMS, now BOEMRE), NOAA and the Department of Defense developed the Multipurpose Marine Cadastre, an integrated information system on submerged lands consisting of legal (property ownership or cadastre), physical, biological, and cultural information in a common reference framework. It identifies outer continental shelf (OCS) locations of federally-permitted activities; obstructions to navigation; submerged cultural resources; undersea cables; offshore aquaculture projects; and areas designated for the purpose of safety, national security, environmental protection, or conservation and management of living marine resources.

Specific efforts are already underway to improve maritime operations. These include:

- Utilizing unmanned vehicles to provide volumetric and bottom sampling data for various industries;
- Improvements in autonomous underwater vehicle design by NOAA and the U.S. Navy;
- Improvement of the integrated ocean observing system and other observing advances, such as implementation of the Ocean Observatories Initiative; and

- Fusing environmental observing and prediction systems with multi-use marine operations prediction and simulation systems to promote economically efficient, environmentally sound, and successful operations across the broad spectrum of marine operations.

There is increased scientific, governmental, and societal attention on the effects of climate change and the need to understand and predict the impacts of change on marine operations. Possible ramifications include necessary changes to ports, harbors, and coastal platforms as sea levels rise and coastal storms escalate in frequency and intensity.

Updated Priorities

Despite this progress, additional attention is needed on the interaction between maritime operations and the environment and predicting conditions in the marine domain.

Research Priority 1: Understand the Interactions between Marine Operations and the Environment

Determining how marine operations (including port development and vessel maintenance and operations) affect the environment, and how environmental conditions affect operations, is central to enabling effective and environmentally sound marine operations.

This understanding will be particularly important in environmentally sensitive areas, such as the Great Lakes, coastal areas with low tidal exchange, and coral reef systems. Areas of study pertaining to effects of marine operations on the environment should include:

- Release, dispersion, cycling, and cumulative ecological impacts of contaminants (e.g., from oil spills and releases, air emissions, non-point sources of pollution);
- Interactions with marine life (e.g., bird migration, impacts of ship strikes and ocean sound on megafauna such as whales);
- Impacts of aquaculture development and production (e.g. increased nutrient loading, invasive species);
- Factors contributing to the introduction and persistence of invasive species (e.g., ballast water);
- Influence of fuel choice on emissions (e.g., black carbon); and
- Effects of expanding transportation routes (e.g., in the Arctic).

These research efforts will build upon investigations into factors that influence ecosystem dynamics (e.g., invasive species, contaminants) and sustainable resource use (e.g., impacts on wild stocks from aqua-culture). They will also enable the development of mitigation strategies and technologies (e.g., passive sonar, spill prevention and containment, ballast-water treatment, vessel routing to avoid ship strikes).

Research is also needed to address the feedback effects of environmental conditions on marine operations. Areas of study should include, for example, refining models of sediment transport to enable rapid, efficient, and environmentally sustainable dredging and dredged-material management. Investigations of environmental changes related to climate change (e.g., sea-level rise, sea-ice abatement, lake-level decreases, ocean acidification) can inform the development of alternative transportation routes, necessary changes to ports and harbors, shifts in fishing patterns, and the design of energy-extraction platforms to maximize stability and safety.

Interactions between marine operations and the environment will vary by region, so the type and geographic scope of sampling will necessarily need to be adjusted by location. Since the volume and diversity of marine operations increases closer to shore, finer-scale modeling is needed there, and is likely more feasible as well, as sensing capabilities are typically denser in the nearshore. The cumulative impact of land-based activities upon the natural environment also renders nearshore areas a priority.

Research Priority 2: Improve Understanding of Environmental Factors Affecting Marine Operations to Better Characterize and Predict Conditions in the Maritime Domain

Better environmental observation, characterization, and forecasting of ocean and waterway conditions (e.g., winds, currents, turbidity, surface waves, sea-ice extent, lake levels, water temperature, biogeochemical conditions) are necessary for safe and efficient marine operations, including transportation, military operations, offshore energy projects, and search-and-rescue efforts. Increased precision in forecasting marine conditions (e.g., improved real-time ocean current models; storm-surge projections),

in conjunction with efforts to address natural hazards, can help minimize negative interaction between marine operations and the environment. Technologies (e.g., robust sensors, autonomous vehicles) should be developed to improve data collection in all weather conditions, to support high-spatial-resolution and near-real-time forecasting throughout the ocean and coastal zone. These observing systems need to be integrated with one another to support more comprehensive models and other tools for decision making. Increased observation capabilities will also facilitate long-term forecasting, allowing maritime operations to anticipate changes that will occur with climate change (e.g. the opening of Arctic sea routes).

Areas of particular focus should be safely expanding Arctic transportation routes and "greening" the marine transportation system. Changing patterns of ice formation will allow for more regular surface transit through the Arctic. Close monitoring and forecasting of atmospheric and ocean conditions such as sea height, wind, ice, and severe storms are necessary to ensure safe transit.

Photo 13. U.S. Coast Guard Cutter Healy breaks ice in the Arctic. Credit: U.S. Coast Guard photo by Petty Officer Patrick Kelley.

Tools to be Emphasized

Achieving safe, effective, efficient, and secure marine operations that support environmental protection requires development of an end-to-end process that moves from data collection to modeling to design of user-support tools. This includes a diverse suite of infrastructure and technology, ranging from assessment methodologies (e.g., rapid-assessment methods for detecting marine contaminants/pollutants, pathogens, and both harmful and non-indigenous species) to observing systems and information networks. Providing accurate environmental information depends upon robust observational networks to monitor, record, and present real-time surface-monitoring data (e.g., high-frequency, coastal-based radar facilities, Vessel Monitoring Systems, Automatic Identification Systems, GPS-based deliveries). In addition to maintaining the continuity of the global ocean observing satellite capability, priorities include:

- Advancing sensor and technology development, particularly for autonomous and persistent observations, as well as for long-term observing systems;
- Real-time or near-real-time data collection on environmental variables by incorporating; observational capabilities of ships of opportunity (e.g., fishing, cargo, and passenger vessels)
- Automated and autonomous bottom-mapping capabilities for detecting changes, to improve scheduling of rapid, full-scale surveys; and

Management of data is as important as its collection. Data collected by the observing systems should be accessible through a comprehensive national data network, either through a single system[31] or a coordinated distributed network. Developing this data network will require new ways to address gaps in data collection, sharing, and interoperability of technologies, and should permit integration of existing research into operational systems (e.g., systems that provide real-time navigation data to vessels). This data network should be able to link with other databases, such as those focusing on ecosystem data, and developed in accordance with international standards for data exchange. If constructed, the national data network will also provide the data needed for models that simulate various scenarios to help us understand potential impacts of weather events or man-made disruptions on marine operations, and to support restoration plans for operations.

Steps have been taken to establish key elements of a data network through several avenues including the National Ocean Policy, ICOOSA, and the CMTS Navigation Technology Coordination and Integration Team. Integration of existing regional observing systems and the building and integration of new sensor networks are envisioned under the National Ocean Policy.

Concurrent with improvements in data collection and management, it is a priority to incorporate data into predictive models. Data become increasingly useful as they are interpreted and applied as decision-making criteria. An appropriate workforce should be trained in model and forecast development, distribution of products to appropriate users, and interpretation of the impact of those products on various sectors of marine operations. Such developments will be particularly challenging in the Arctic as communication infrastructure is notably weak in that region and there is a paucity of ocean weather observation stations.

A well-trained workforce – along with an educated public – is a priority for facilitating coordination of marine operations. This workforce will be needed both in the lab, developing models and forecasting tools, as well as on the water, working in various marine operation sectors. Frequently marine uses are isolated from one another, with users in one sector relatively unaware of needs and operations in another. For the most effective marine planning, it is a priority to close this information gap. In addition, greater emphasis needs to be placed upon including the public in education efforts, especially concerning how their actions affect marine operations and the environment.

Regional Applicability

Dominant considerations regarding marine operations will vary by region, and thus, the environmental, technical, and social data that are relevant will likely differ. For example, on the Gulf Coast, monitoring of oil discharge and fate might be emphasized while in the Northeast, commercial shipping impacts could be a primary focus. The impacts of climate change upon marine operations also vary by region, so modeling should be tailored by region. This is particularly relevant to the issue of sea-level rise; dramatic loss of land is being seen in low-lying and subsiding coastal areas. Both scenarios have the potential to significantly impact ports and inshore marine operations.

While there is an emphasis upon the collection of physical science data needed to conduct marine operations, knowledge gathered through the social sciences regarding economics and human use, conflict, and perceptions is also necessary to avoid conflict between various marine operations and to understand human use patterns. For example, information about the location and seasonality of fishing practices is important when considering the placement of offshore energy projects. Collaboration across sectors is also vitally important.

The environmental data collected and processed by academic researchers should be available to meet the needs of industry and the legal regulations of local, tribal, state, and Federal governments. A successful demonstration of this type of collaboration was seen in the 2007 shifting of the Boston traffic separation scheme to reduce whale strikes in Stellwagen Bank National Marine Sanctuary.

Theme 4: The Ocean's Role in Climate

The global ocean plays a fundamental role in governing the planet's climate through its capacity to store and distribute heat, its major role in the carbon cycle (including storing CO_2 from the atmosphere), its exchange of trace gases with the atmosphere, and its role as both source and sink of some atmospheric particulates. Our scientific challenge is to accurately assess the ocean's past and present state, processes, and phenomena that influence climate, and society's influence on them. Predictions and projections of climate variability and change over a range of spatial scales from global to regional and local, and at timescales from centennial to interannual and intra-seasonal, are necessary. These predictions and projections will improve society's ability to respond to and reduce, where feasible, climate-related hazards; to adapt to global climate change and variation (e.g., sea-level rise, changing weather patterns); and to inform management and policy decisions addressing human and environmental impacts. Indeed, improvements in our predictive capability and understanding of ocean climate will improve our ability to deal with all of the other themes discussed in this report.

Rationale

The ocean covers more than 70 percent of the planet, and it is Earth's second largest reservoir of anthropogenic CO_2. It absorbs a significant amount of the excess gas emitted into the atmosphere. In doing so the ocean buffers our climate from the greater temperature increases that would occur if all the added CO_2 remained in the atmosphere. Potential changes in the rate at which the ocean absorbs this added CO_2 could have significant impacts on future global climate change. Changes in ocean conditions have been demonstrated to affect marine ecosystems, causing, for example, large changes in marine populations that support fisheries, changes in seabird population distributions, and an increase in coral-reef bleaching events. The quantity of greenhouse gases, such as CO_2 and methane, in our future atmosphere will depend, in part, on the exchange of these gases in ocean and coastal systems (e.g., wetlands). Rising atmospheric CO_2 levels lead to increased uptake of this gas by ocean waters, resulting in ocean acidification. Substantial changes in ocean chemistry will further influence ecosystems and their health. A more acidic ocean leads to decalcification of carbonate organisms, such as corals, microscopic plants, and animals critical for supporting life both in the ocean and on land. These pH changes are taking place along with temperature changes, and the connections between the two should be understood. The growing body of knowledge about the impact of climate on marine chemistry and ecosystems will be critical for informing ecosystem-based management efforts.

The ocean is also a great storage facility for the planet's heat. To date, imbalances in the planetary energy budget are manifested in changing ice conditions, sea level, and ocean biogeochemical composition, ecology, and temperatures. Rising sea level as a result of runoff from melting ice sheets and glaciers and also of expanding ocean volume due to higher temperatures is a major threat to coastal communities. Over the past few decades, global sea level has risen faster than before, resulting in a challenge for climate modeling. Ocean circulation dynamics cause uneven and relatively unknown geographical variation, with sea level rising and falling at different locations. In addition to the obvious threat from episodic events (e.g., storm surges, coastal flooding), low-lying coastal regions are particularly impacted by gradual sea-level rise. A more complete understanding of how changing sea level impacts coastal communities and ecosystems is needed at regional, state, and local levels. The importance of this understanding is increasing as the already large fraction of the world's population living in coastal regions continues to grow. Given the particular vulnerability of coastal residents of developing countries, climate-change-induced sea-level rise is an issue with enormous potential for human impact and thus has implications for both humanitarian and national security concerns.

The polar oceans are under a particularly large threat from climate change. As demonstrated by the decrease in extent and thickness of Arctic sea ice over the past 30 years, they are changing rapidly. These decreases will, in turn, accelerate global warming, as highly reflective ice is replaced by much less reflective liquid water. Ocean biology also responds to the change in sea ice distribution, as photosynthetic productivity can take place in regions where it was previously limited by opaque ice cover. More importantly, increased stratification of high-latitude waters due to increased temperature and incoming solar radiation can lead to major shifts in high latitude ecology, shifts from benthic- to pelagic-based food webs, and broad habitat changes. These can have a cascading effect upon global fisheries.

The ocean is important in other planet-regulation processes as well. The tropical ocean is a vital component of climate variability, at scales from seasonal to interannual. For example, sea-surface temperature variability associated with the El Niño and La Niña phenomena has a substantial impact on many regions of the world, influencing agricultural yields, hurricane intensity, and droughts. The global ocean is also a significant component of the planetary water cycle. It receives and redistributes freshwater from rivers and ice discharge, and provides moisture (through evaporation) to the atmosphere that precipitates over the ocean and continents. Alterations in large-scale ocean circulation can influence long-term global climate change, and possibly short-term or abrupt changes.

The ocean also has a controlling influence on the path and intensity of major storm systems, such as hurricanes, mid-latitude winter storms, and intraseasonal atmospheric oscillations. Improved understanding of the drivers of these systems, resulting in improved predictive capability, will help society to prepare and adapt cities and other public infrastructure for the inevitable arrival of severe events as well as to take advantage of opportunities presented by climate change.

Ocean phenomena that contribute to global climate change and variability include large-scale, longterm coherent processes such as the Pacific Decadal Oscillation (PDO), North Atlantic Oscillation (NAO), and Atlantic Meridional Overturning Circulation (AMOC); large-scale, nonlinear behavior (e.g., abrupt changes in large-scale ocean circulation); and ocean–atmosphere chemical and biogeochemical fluxes and process interactions (of carbon, momentum, heat, trace gases, and aerosol precursors).

A broad understanding of the ways that major ocean ecosystems interact with climate through these phenomena should be developed so as to include this knowledge in climate models. Also, applying improved understanding of the ocean's past and present role in global climate change and variability will enable better predictions and projections of climate effects on ocean processes and components (e.g., species, ecosystems), as well as of ecosystems'feedbacks to climate variability and change. This will help to inform current and future management efforts, including coastal land use and development, and alternative energy use and development (e.g., wind and tidal power).

The National Ocean Policy highlights the importance of understanding and adapting to climate change, one of nine priority objectives. This objective specifically singles out the impact of ocean acidification as an area of particular focus in research and adaption planning efforts. The Arctic ecosystem is also noted as being particularly vulnerable and therefore requiring prioritized attention.

The ocean's role in regulating climate and the impact of increasing CO_2 on the ocean will be necessary components in implementing the National Ocean Policy.

Research Priorities

Progress to Date

In terms of advancing understanding, the National Science Foundation has implemented a number of climate-related research programs. Topics include water sustainability, ocean acidification, biodiversity, and climate prediction. Investments are also being made in partnerships for climate change education.

It has only recently become possible to map ocean biomes (major types of ecological community) from space in near-real-time and synthesize causes of human-induced change and human impact on over 50 marine ecosystems. This provides the basis for detecting and quantifying changes in ecosystems resulting from human impacts and climate change.[32] There is also new documentation of acceleration in sea-level rise, and newly-improved understanding of the components of sea-level rise, the way the ocean functions as a carbon sink and its implications for ocean acidification, the sensitivity of Arctic Sea ice to small changes in climate, and the limitations of current models. The Department of the Interior has established climate adaptation centers and regional landscape cooperatives to better understand and coordinate efforts to adapt living resources and ecosystems to climate change.

Research on the Atlantic Meridional Overturning Circulation (AMOC), a near-term priority of Charting the Course, has yielded progress on understanding the role of AMOC in global climate. Recent studies indicate that temperature variations in the North Atlantic on decadal to multidecadal scales are caused by variations in AMOC. In addition, modeling indicates that an intensifying AMOC, in addition to anthropogenic greenhouse-gas–induced global warming, may have contributed to the recent decline in Arctic sea ice. There is also increasing evidence that changes in the upper, warm limb of the AMOC are affecting the mass of the Greenland ice sheet.[33]

Consistent with the scientific challenges identified by the USGCRP, it is essential to improve understanding of the ocean's role in past, present, and future global climate and to educate decision makers and the public about this role. As the USGCRP evolves, there will be improved focus on adaptation, vulnerability, impacts, and mitigation. Developing these areas will require synergy between ocean research and the broader climate program.

Offering the public more information, including regarding how individual actions cumulatively affect the oceans and climate, will improve ocean literacy and ideally result in behavior modification. Steps taken to improve ocean literacy as it relates to climate change include establishment of the

Communication and Education Program of NOAA's Climate Program Office and the newly launched Climate Service portal.

Updated Priorities

Research Priority 1: Understand Ocean–Climate Interactions within and Across Regions

Ocean regions, such as tropical, polar, and deep sea, can exert powerful influences on, and in turn be influenced by, global climate change. Current understanding of the type or temporal and spatial extent of this influence is limited. Improved short-term climate predictions require a more complete understanding of the influence of global tropical ocean phenomena (as demonstrated by El Niño, AMOC, PDO, and NAO events, and monsoons in the Indian and eastern Pacific Oceans). Attention to changes in the Arctic is especially critical. Increasing global temperatures will eventually lead to an ice-free Arctic Ocean in summer, with potentially widespread impacts such as changes in polar albedo (reflective power) and ocean–atmosphere heat exchange, alterations in sensitive Arctic ecosystems, and development of new shipping routes (see Theme 3 Maritime Operations and the Marine Environment). Recent data indicate that these impacts are materializing more rapidly than imagined just a few years ago. A warmer Arctic can contribute to coastal flooding due to melting ice sheets, and may influence abrupt or longer-term climate change. The Southern Ocean, one of the most poorly observed and hostile environments of the Earth, is a region of significant biological productivity and CO_2 uptake, and is influenced by large-scale ocean circulation with significant climate implications. Finally, the role of the deep ocean should be ascertained, particularly with regard to the ways it mitigates climate change (e.g., via carbon sequestration and heat storage). Processes within these regions (e.g., ocean circulation, air–sea interaction, convection, water-mass formation) can have a variety of individual and compounded impacts over many spatial and temporal scales, ranging from the regional effects of diminishing sea ice on Arctic ecosystems to global effects of sea-level rise.

Research Priority 2: Understand the Impact of Climate Variability and Change on the Biogeochemistry of the Ocean and the Implications for Ocean Ecosystems

Changes in the ocean's physical properties (e.g., heat, circulation) and biogeochemical properties (e.g., carbon, nitrogen, dust, trace elements, pollutants) can have a variety of impacts on ecosystems ranging from coastal watersheds to shallow- and deep-water coral reefs to open-ocean systems. Of particular concern is the impact of ocean acidification on ecosystem function. In addition, altered biogeochemistry, changing current patterns, loss of sea ice, rising sea levels, and modified salinity and sea-surface temperature may irreversibly alter ecosystems. Sustained observations (e.g., from global and coastal observatories), research on processes (e.g., air–sea exchange, ecosystem interactions), and modeling (e.g., ocean–atmosphere–land models integrating global and regional biological, chemical, and physical data) will help determine fluxes and cycling of biogeochemical variables and help identify and quantify impacts on ecosystems of the greatest importance. These efforts will also help identify subsequent feedbacks regarding the influence of ecosystems on climate (e.g., the influence of ocean acidification on carbon uptake and sequestration, desertification, increasing iron dust and subsequent oceanic algal blooms). The knowledge gained, in conjunction with understanding of impacts to ecosystem health, will support effective management of these ecosystems so that they remain healthy and viable.

Research Priority 3: Apply Understanding of the Oceans to Help Project Future Global Climate Changes and Their Impacts

The oceans are getting warmer, more freshwater is being added by melting ice sheets, and more CO_2 is being absorbed from the atmosphere. Melting sea ice provides the potential for a significant feedback for high-latitude warming, as well as affecting ocean productivity, leading to changes in high-latitude oceanic ecosystems. The ability to predict the timing and magnitude of regional climate change due to these and other processes, and to understand how such change will impact society, will be improved by quantifying the predictability of these processes and by increasing the temporal and spatial resolution and dynamic complexity of global ocean–atmosphere–land models. Integrating a robust set of global and regional ocean observations, coordinated paleoceanographic data and assessments, and the results of improved process research into such models will help determine the past and present influence of ocean processes on climate change, including the potential for rapid or abrupt change (at "tipping points"). Coupled climate models will support improved short-term predictions (e.g., hurricane intensity) and long-term projections (e.g., sea-level rise) that will enable local, tribal, state, regional, and Federal policy and decision makers to plan for and adapt to effects of climate change. The science community's overall interest in "seamless weather-to-climate prediction" suggests that future models should have more accurate representation of the key processes that couple the ocean and atmosphere.

Tools to be Emphasized

A continuing challenge for U.S. climate-related ocean research is the lack of a robust, integrated system of global and coastal ocean observation, including *in situ* and space-based sensors, that is sustainable and connected with international observing efforts. Particular emphasis should go to establishing the validity of remote sensing methods in coastal regions and other areas of high complexity and gradients. In addition to collecting physical, biological, and biogeochemical data, the observing system effort should be able to integrate new technologies and incorporate new contributions without compromising climate records (which will become increasingly important as more countries contribute in a significant way to observational databases). Incorporating proxy ocean data (e.g., paleoceanographic data from sediment and ice cores) into the larger ocean–climate information system is also needed. Such an information system should have the capability to develop climate-data records for physical, biogeochemical, and ecological datasets; to reconstruct past ocean states; and to assist in developing and refining climate proxies.

Another barrier to improved understanding is insufficient knowledge of how biological, chemical, and physical ocean processes interact through a variety of space and time scales. Obtaining this knowledge will require that ocean field campaigns become increasingly interdisciplinary, combining physical and biological/biogeochemical research far more intimately than in the past. Ship-based research should also more fully incorporate global *in situ* and space-based measurements and modeling as a context for field campaigns, and allow results to be applied generally. Ship-based research should utilize newly capable platforms, such as gliders and floats.

Improvements in ocean and coupled climate model and data assimilation will be necessary to integrate large volumes of observational data, reconstruct past ocean states, and predict future states. Increased computational capabilities, including versatile software, efficient algorithms, focused resolution, and capacity, are required to support these advances in data and modeling capabilities and allow for providing results to the research and relevant operational communities. An end-to-end system in which observations feed modeling, which in turn produces forecasts for use as input to decision tools, is a necessity.

Sustained and improved satellite and *in situ* sensors are necessary to collect information over a sufficient time period so they can detect subtle background climate trends in a broader range of climate parameters, such as currents, salinity, and sea-ice thickness, to draw a more complete picture of fundamental climate processes. Next-generation, *in situ* chemical and biological sensors that collect a variety of information, including data on sentinel organisms and habitats, will also be important. Improvements in data collection should be accomplished while maintaining long-term climate records of key variables. Advances in methods of ocean data assimilation are required to allow for true integration of *in situ* and remotely sensed data, especially satellite-based column and surface measurements and *in situ* vertical profile measurements.

The scientific community required to address the ocean–climate connection will thus need to be more interdisciplinary than ever before, and able to work with a variety of data sources (both *in situ* and remote sensing), utilize models for prediction and/or dataset production, and work effectively in a far more international environment than ever before. This will place significant burdens on the training of scientific personnel in this area, given the need for additional breadth and ability to focus as part of a large team.

In addition, the social sciences are an important source of information for describing ocean–climate interaction and anticipating changes such as location and importance of local fishing communities as resource locations shift with climate change; and offer social, behavioral, and economic data, as well as supporting local ecological knowledge. Such knowledge can be especially important in remote areas and can provide vital facts about ecosystem changes. It is vital that these commonly untapped resources be utilized in order to capture a more complete picture of climate change.

Regional Applicability

The effects of climate change are not uniform across space, and therefore planning to address its impacts should necessarily be tailored to specific regions. Sea-level change is a particularly poignant example of regional variability. Sea-level rise in low-lying coastal areas have made land loss a significant problem on the Gulf Coast. Differences also occur with fish stocks, where the range of certain species is shifting due to changing ocean conditions. Changes in the Arctic have particularly significant implications for changes in ocean circulation, greenhouse gas accumulation, planetary warming, and human activity.

The large areas involved, and the multitude of agencies with regulatory authority over the ocean, our coasts and coastal watersheds, and the Great Lakes, mean that coordination among research institutions and management entities is essential to success. Regional bodies have already formed across the coastal states to address ocean issues in a more holistic fashion. Coordination with international partners is also crucial, since climate change is a global phenomenon, unconstrained by national boundaries.

Theme 5: Improving Ecosystem Health

Human society reaps enormous and diverse benefits from the complex marine ecosystems that exist off our shores. Food production, clean drinking water, climate regulation, intangible aesthetic value, and protection of life and property from storm damage are only a few of the many benefits provided by the ocean, coastal, coastal watershed, and Great Lakes areas. In the past century the impact of humans on ecosystems has intensified. As human populations grow, they draw more and more on natural resources to produce net short-term gains at the expense of biodiversity and ecosystem services. Diminished water and habitat quality, enlarged dead zones, new water-borne diseases, decline in habitat quality, and fisheries collapse are a few of the effects of reduced marine and freshwater ecosystem health and the resulting degradation in diverse ecological services. Addressing the cumulative effects of human impacts in order to minimize and reverse detrimental effects will require a more holistic, ecosystem-based approach to management.[34]

Rationale

The services provided by an ecosystem are inextricably linked to its overall health. Degradation of ecosystem health can have non-linear and unforeseen consequences that current management actions may not be able to mitigate. Accurate valuation of the full range of provisioning, regulating, cultural, and supporting services provided by natural ecosystems is a key component necessary for consideration in the management of human activities at the coasts and in the oceans. Provisioning services include the goods supplied by coastal and marine ecosystems, such as fishery catches, oil and gas, renewable energy, and food from aquaculture. Regulating services include regulation of climate and water quality, storm mitigation, and pollution absorption. Cultural services include recreational, spiritual, aesthetic, educational, and cultural heritage activities; supporting services include carbon sequestration and primary production. To effectively improve ecosystem health there should be a rigorous scientific understanding of the structure and function of marine ecosystems that are responsible for providing ecosystem services and an ability to apply that understanding to develop a comprehensive ecosystem-based management framework.

Climate change and its many ramifications, such as ocean acidification and sea-level rise, can potentially have extreme, abrupt, and/or irreversible impacts on ecosystems. In general, ecosystem degradation happens slowly and somewhat predictably, resulting in a gradual loss of services rather than rapid, and therefore more noticeable, changes in ecosystem structure and function. However, it is possible that ecosystems may degrade slowly through a number of mechanisms but then reach a tipping point at which a sudden change occurs, moving the ecosystem into a different, semi-permanent state. Though our capability to predict change is improving, predicting the thresholds at which cascade effects can occur is still very difficult.

The National Ocean Policy includes a precautionary management approach when, for example, existing information is insufficient to address second-order effects of interactions among ecosystem components. Robust decision-making will result when decisions are informed by timely, specific, and pertinent information. While much remains to be learned, much can be accomplished with the present level of knowledge to proactively conserve and improve ecosystem health, thereby ensuring the delivery of ecosystem services.

To reverse declining ecosystem health, the National Ocean Policy calls for integrated conservation and management of ecosystems through regional marine planning. The emphasis on ecosystem-based management and marine planning implies a supporting suite of observation, research, and assessment activities. The research priorities outlined in this section seek to address some of the highest priorities in assessing and addressing ecosystem health concerns and helping to support valuation of ecosystem services.

Research Priorities

Progress to Date

The Comparative Analysis of Marine Ecosystem Organization (CAMEO), named by Charting the Course as a Near-term Priority, is playing an important role in increasing our understanding of ecosystem structure and function, by enabling the gathering and integration of data related to ecosystem production and organization at the regional level. For example, the Puget Sound

Partnership in Washington State is using ecosystem health metrics and modeling to provide stakeholders and managers with a framework for decisions. The Partnership uses Integrated Ecosystem Assessments (IEAs) as a tool to help identify what ecosystem restoration activities will have the greatest impact on the improvement of ecosystem services. An IEA both integrates and analyzes diverse scientific information, and requires cooperation across a spectrum of partners ranging from Federal Government to state and local governments as well as academic, non-governmental organizations, and industry partners. The goal of an IEA is to inform decisions that support an ecosystem-based approach to managing marine resources and achieve environmental, economic, and societal objectives. An IEA is dynamic, iterative, and adaptive. The IEA program for the California Current is working to develop visualization tools designed to provide easy access to multiple sources of environmental, economic, and social science data. The program works across Federal, State, and local agencies as well as with stakeholders to address pressing coastal and marine management goals in that large marine ecosystem. This effort has achieved better understanding among the whole spectrum of stakeholders of how ecosystem-based management should assess ecosystem services, as well as the methods of valuation that underpin those assessments. IEAs also provide the basis for an adaptive approach to management, which acknowledges uncertainties in how complex ecosystems will respond to changes in human activities.

Updated Priorities

Research priorities for improving ecosystem health should focus on the most important uncertainties, and work to fill gaps in knowledge most needed for both near-term decision-making and long-term understanding. The inclusion of current research results is essential for improving the effectiveness and adaptability of ecosystem-based management.

Research Priority 1: Understand and Predict the Impact of Natural and Anthropogenic Processes on Ecosystems

The complex relationships that determine ecosystem structure and function vary in scale both spatially (e.g., local, regional, basin-wide, global) and temporally (e.g., seasonal, annual, decadal, centennial), and incorporate various feedback effects. Shifts in ecosystem structure can be induced by climate–ecosystem interactions (e.g., impacts of El Niño/La Niña, increases in ocean temperature), human activities (e.g., watershed activities, sediment/nutrient/contaminant flux, fishing), and productivity-driven or predation-driven trophic mechanisms (e.g., invasive species). Among these, increasing ocean acidity is an especially critical issue in need of focused research. Understanding climate–ecosystem relationships requires incorporating new and existing knowledge and observations at appropriate temporal and spatial scales to explore the full range of physical, chemical, biological, and ecological mechanisms that determine observed ecosystem structure, function, and variation. Improved ecosystem decision-making requires:

- Comprehensive analyses of natural and anthropogenic changes in physical, biological, and chemical properties and their individual and cumulative impacts on productivity and overall ecosystem health;
- Assessment of dispersal mechanisms for marine organisms;
- Development of next-generation trophic dynamics models, spanning multiple trophic levels; and
- Process studies and model development to assess impact (loss) and recovery responses to natural and anthropogenic declines. For example, linking the extent and effects of the hypoxic zones in the northern Gulf of Mexico to land-use practices in the Mississippi River drainage requires complex models.

Research Priority 2: Apply Understanding of Natural and Anthropogenic Processes to Develop Socioeconomic Assessments and Models to Evaluate the Impact of Multiple Human Uses on Ecosystems

Understanding human impacts on marine ecosystems, whether positive (restoration) or negative (degradation), will require integrating traditional ocean science with socioeconomic science. This is especially true in the Arctic because of the rapid changes in this region. Restoration science as a body of principles and best practices is rapidly growing in importance and relevance to our Nation's coastal resilience and sustainability.[35,36] To determine and predict society's impact on marine ecosystems, social and economic factors (e.g., land, water, and energy use; coastal and watershed development; resource-use perception; cultural history) that determine how society views and uses marine ecosystems should be assessed and modeled. Existing non-market valuations of ecosystem services and processes are often geographically and topically fragmented, making temporal and spatial analysis difficult. Efforts should include developing new methods for evaluating non-consumptive use of ecosystem services and characterizing the values society places on competing uses. To ensure the sustainability of ecosystem goods and services, those methods should consider the rights of future generations and include a discount rate for adjusting cost-benefit analyses over time.

Photo 14. Beach clean up at the Pacific Missile Range Facility, Barking Sands, Hawaii. Credit: U.S. Navy photo by Mass Communication Specialist 1st Class Jay C. Pugh/Released.

Research Priority 3: Apply Understanding of Marine Ecosystems to Develop Appropriate Indicators and Metrics for Effective Management and Sustainable Use

A robust suite of indicators of ecosystem structure, function, productivity, and services should be used and evaluated and implemented at multiple scales (local, regional, basin-wide). These indicators will help in assessing factors that stress and degrade ecosystems (e.g., ocean acidification, eutrophication and harmful algal blooms; loss of coastal wetlands; shoreline development; overuse of harvested species; invasive species; introduction and cycling of contaminants; changes in biodiversity, ecosystem productivity, and resilience). Additionally, indicators and metrics are needed to help monitor the restoration and recovery of degraded ecosystems. Input from a range of scientific disciplines is needed to identify such indicators, or to develop them where existing ones are insufficient or where gaps are identified. Metrics and indicators should evolve based on new scientific understanding and management needs and goals and be evaluated through integrated ecosystem assessments. In this manner they will provide feedback for assessing management efficacy and a basis for incorporating new information and understanding to adapt and improve management practices. The development of ecosystem approaches to marine planning will require four-dimensional models of ecosystem response to alternative use scenarios. This type of system will allow projections of climate effects on long term changes in ecosystem productivity and the geographic extent of ramifications of decisions or lack thereof. Planning decisions will require understanding of the sources and sinks (fate) of marine animal larvae, valuation tools to assess tradeoffs among various potential use sectors, and visualization tools to comprehend the likely impacts of planning decisions on ecosystems.

Tools to be Emphasized

Effective observing capabilities should provide essential physical, chemical, and biological data on various ecosystem types ranging from terrestrial watersheds to productive coastal and continental shelf regions to the deep, pelagic realms. Collection of such data will require extensive infrastructure including research vessels, automated buoys, and autonomous vehicles for short- and long-term sampling of water-column properties; satellite-based assessment of surface characteristics (e.g., temperature, biogeochemical properties, ocean color, surface currents, wave heights); *in situ* observatories in the ocean and across the land–water interface; shore-based facilities for sample analysis, experimental manipulation, and observing system maintenance; and a range of survey (e.g., mapping) capabilities.

Improvements in, and maintenance of, information technology and infrastructure are essential to ensure that data assimilation, analysis, and modeling tools are available to accommodate and enable the integration of ecosystem and socioeconomic data. For example, ocean color is an important ocean parameter, which contributes to information on a variety of ocean-related issues, including sea surface temperature and ocean acidification. Maintaining capabilities for satellite observations of ocean color and other parameters is a priority, even when national and international commitments to satellite coverage are uncertain.

Understanding the causality of observed changes in ecosystems informs an adaptive management approach. Implementing adaptive management measures in response to observed changes will improve success and enable stronger linkages between science and the beneficial social outcomes of ecosystem-based management. It is critical that we develop end-to-end systems, based on collaboration between academic and operational agencies, to develop tools (such as the IEA program) to support ecosystem decision relative to local and regional issues, such as nutrient management, fisheries management, and marine planning. Four-dimensional visualization tools are needed to provide geospatial information supporting decisions based in science and a thorough understanding of the "downstream" consequences of action or inaction. Such tools are data-intense, with particular emphasis on time series as key to model validation. Effective management of vast amount of data from multiple sources and with variable levels of sophistication is a key element in the inter-operation and functionality of ecosystem-based models.

Fundamental to all research progress is human capital. Ensuring an effective and adaptive ecosystem-based research and management approach requires an investment not only in technology and infrastructure, but also in education systems to produce the scientists and managers needed to implement this approach. Collecting and analyzing data, conducting monitoring and assessments, and observing changes in ecosystems are all tasks that should be accomplished by a dedicated and knowledgeable workforce. Ecosystem research requires collaboration between aquatic and terrestrial natural sciences (e.g., biogeochemistry, taxonomy, systematics[37]) and social sciences (e.g., sociology, economics) as well as individuals who can transform data into information products for end users.

Regional Applicability

The impact of humans on ecosystems is mostly at local to regional scales. Each region faces different anthropogenic and natural stressors that should be considered at the appropriate scale. Scaling models to the appropriate levels will be central to assessing and forecasting ecosystem health. In many situations ecosystem models will need to be scaled up so they are portable to other geographic areas, while climate models will need to be scaled down to improve regional forecasting abilities. The consequences of climate change will vary by region, necessitating regional approaches to management (e.g. melting Artic sea ice, ocean acidification, sea-level rise). The National Ocean Policy proposes nine ecosystem planning regions that provide a framework for addressing scale in the implementation of ecosystem-based management. Within each region there is likely to be a hierarchy of smaller-scale ecosystem-management issues.

Ecosystem health also has global implications, as an ecosystem impaired in one country can have downstream impacts on others. Ecosystems are not confined by geopolitical boundaries, but they often are intersected by jurisdictional lines which may dictate management action or inaction. Challenges to recovering and sustaining healthy ecosystems in developing countries are different than in similar recovery efforts in the United States, possibly requiring different techniques and solutions; collaboration among stakeholders across boundaries is often critical.

The many programs organized around the concept of Large Marine Ecosystems (LME), many coordinated through the World Bank and Global Environment Facility, offer an example of successful establishment of science-based conservation and management strategies at regional scale around the world. The LME programs build technical capacity and allow countries to take ownership of their successes and failures relative to ecosystem health. In the long run, healthy ecosystems that deliver necessary services contribute to political stability as well as human health.

Photo 15. Harvesting prohibitions due to health threats in Maine. Credit: National Ocean Service, National Oceanic and Atmospheric Administration.

Theme 6: Enhancing Human Health

The ocean, our coasts and coastal watersheds, and the Great Lakes hold vast resources that convey health benefits by providing food, oxygen, waste recycling, climate moderation, carbon sequestration, recreation, and energy. In addition, exploration of extreme habitats and the discovery of new species and novel chemicals, combined with emerging biochemical and biotechnical techniques, promote the development of bioproducts that promote human health. Environmental change puts these benefits at risk through the depletion of ocean and coastal resources, degradation of environmental quality, the loss of ecosystem services, and increased threats to human health and livelihoods.[38] Direct health hazards include catastrophic hurricanes, flooding, tsunamis, contaminated seafood, polluted waters, disease-causing microbes, and harmful algal blooms (HABs). Understanding the causes of health hazards and how they can be mitigated or managed can decrease the numbers of resulting illnesses. In the next decade, efforts to improve our understanding of oceans and human health will focus on explaining the causes and emergence of diseases and new techniques and products to prevent and treat disease. Enhancing human health will ultimately depend on translating research results into policies that reverse trends in ecosystem degradation and curb losses in biodiversity.

Rationale

The ocean, our coasts and coastal watersheds, and the Great Lakes present health risks to people through the consumption of water or seafood containing pathogens, toxins, or chemical contaminants; direct contact through recreation, work, and weather events; and indirect contact (e.g., breathing salt air that contains algal toxin aerosols). The frequency and distribution of marine HABs is increasing[39] in U.S. waters and worldwide. Every coastal state in the U.S. has experienced HABs over the last several years, and new species have been found in locations not previously known to have problems.40 The most severe impact to human health is shellfish and finfish poisoning from consuming seafood contaminated with HAB toxins. Public warnings of threats from swimming, eating seafood, and playing in beach sand help curb health impacts, but illnesses often have complex, poorly understood interrelationships with land use practices, water quality, nutrient concentrations, and ecosystem degradation. The emphasis on seafood safety associated with the Deepwater BP oil spill illustrates the risk to the food supply from catastrophic pollution events and the need for robust monitoring and research. Climate change has the potential to cause additional stress on already degraded habitats and at-risk species and change the distribution of microbes, parasites, and invasive species, including some that cause disease. Increasing temperatures, sea-level rise, and ocean acidification are predicted to have major impacts on the structure and function of ecosystems and thereby alter where people can live and work and what they can eat.

Effectively managed and protected, the oceans will provide resources for medical advances. Marine microbes, plants, and animals are sources of thousands of unique chemicals with the potential to treat human diseases. In recent decades, scientists have discovered ecological communities in the ocean with unique biochemical systems, such as those associated with thermal vents and hydrocarbon seeps. These communities, along with ones yet to be discovered, hold huge possibilities for studying biochemical adaptations, pathogen resistance, and the control of disease using genomics. Prior discoveries that now have practical application include pharmaceuticals, diagnostics, molecular probes, and nutritional supplements.[41] Substantial economic benefits have been realized from such "biodiscovery."

Research Priorities

Progress to Date

Efforts in recent years to support research in the ocean sciences, biomedical sciences, and management communities have been stimulated by a joint program of the National Science Foundation and National Institute of Environmental Health Sciences to fund four Centers for Oceans and Human Health. Also, NOAA's Oceans and Human Health Initiative has funded three Centers of Excellence and oversees an external grants program to develop multi-institutional research partnerships with academia, other Federal and state agencies, and the private sector. In addition, the SOST *Interagency Working Group on Harmful Algal Blooms, Hypoxia and Human Health (IWG-4H) released the reports Interagency Oceans and Human Health Research Implementation Plan: A Prescription for the Future (2007) and Interagency Oceans and Human Health Annual Report Fiscal Years: 2007–2008 (2009).*

Member Federal agencies are making significant progress with:

- Better detection methods and forecasting capabilities for pathogens and toxins;
- Improved surveillance for HAB-related diseases;

- Improved sensors for rapid identification and quantification of HAB species and development of early warning systems;
- More thorough monitoring of chemical contaminants, including pharmaceuticals;
- Satellite surveillance that improves risk assessments relating to disease-causing microbes;
- Understanding the biogeochemistry of mercury methylation in freshwater and marine environments;
- Techniques that use seabirds, marine mammals, and other marine organisms as sentinels of health threats;
- Beach forecast models that predict recreational beach conditions for the coasts and Great Lakes;
- Testing of new marine-based pharmaceuticals and natural products, including some that hold promise as potential pain medications, skin-care additives, and drugs to treat cancer, AIDS, and Alzheimer's; and
- Improved educational and outreach materials, made available through public Web sites and local Centers for Ocean Sciences Education Excellence (COSEE).

Updated Priorities

Many challenges still exist in understanding the relationship between the ocean and human health. Research needs are multidisciplinary, and studies of human health are strongly dependent on interfacing with complementary research priorities involving stewardship, ecosystem health, climate change, and natural hazards. Substantive improvements in human health will require a more coherent strategy of targeted monitoring with strong linkages to ocean observing systems, advanced sensors, process studies leading to better information dissemination and decision support, and improved disease surveillance to identify trends in disease occurrence (e.g., the Harmful Algal Bloom-related Illness Surveillance System, HABISS). We should prioritize increasing our capacity to assimilate and integrate available data and promote modeling and forecasting that will reduce the uncertainties associated with the occurrence and spread of disease under alternative climate and management scenarios. These efforts should be conducted in concert with multiple partners and stakeholders, including international groups, working collaboratively to resolve mutually identified priorities. We should emphasize better ocean literacy and outreach to inform people about the benefits of clean oceans and coasts and high quality, healthful seafood, the threats of pollutants and contaminants, and the risks of eating fish and shellfish that are tainted with harmful chemicals, viruses, and bacteria. The updated research priorities below incorporate information and proposed actions from the two IWG-4-H reports mentioned above.

Photo 16. King crab being inspected. Credit: National Marine Fisheries Service, National Oceanic and Atmospheric Administration.

Research Priority 1: Understand Sources and Processes Contributing to Ocean-Related Risks to Human Health

Climate change and its impacts have the potential to become the dominant driver of biodiversity loss and changes in ecosystem services. Climate change affects oceans by warming waters, melting sea and land ice (thereby reducing ocean salinity), increasing sea level, increasing ocean acidification, and changing the availability of metals and other essential nutrients required by microbes, plants, and animals. In addition, changes in precipitation, alteration of inland uses, and increased demand for water to serve public and industrial needs may, in some regions, reduce the availability of surface water and groundwater and ultimately impact coastal watershed, estuarine, and coastal ecosystems. Climate change is affecting the Arctic environment more rapidly than other regions. Hence, an important subset of coordinated ocean, climate, and health research should specifically address the Arctic region. Climate change is likely to result in rapid increases in ocean-related human health threats in the Arctic, and these may be exacerbated by social and economic effects of climate and ocean interactions. Assessing changing health risks for Arctic communities and ecosystems will also help us understand, anticipate, and adapt to future changes in other regions.

Robust observations and improved monitoring systems are necessary to rapidly detect the sources, track the spread, and explain the biogeochemical cycling of current and emerging pathogens, toxins, and contaminants in coastal watersheds, shorelines, oceans, and Great Lakes environments. Improved disease surveillance efforts will help to explain current and emerging health threats from exposure to contaminated water, sediments, air, and seafood. Increased knowledge of species interactions and food webs can help track pathogens and chemical contaminants (e.g., through disease transmission, bioaccumulation, biotransformation, biomagnification). The flux of water, chemicals, and microbes through coastal watersheds, and mechanisms that stimulate, transport, and sustain HABs and contaminants, should also be determined. Identifying, developing, and implementing new and improved rapid assays and models, in conjunction with disease surveillance, will enable assessment, monitoring, and prediction of human health risks. Expanded methods and new sensors will also provide crucial information related to the health of ocean, coasts, and Great Lakes wildlife populations and their role as environmental sentinels and biomedical models for human conditions. The resulting improved understanding will enable development of mitigation strategies and technologies for managing stormwater runoff and wastewater discharge, and lead to new approaches to minimizing the health-related impacts of hurricanes and other natural hazards.

A new effort is needed to combine current and anticipated advances in detection, indicators, and warning systems for ocean and human health threats with actual data on disease occurrence. A Health Early Warning System (HEWS) approach would better provide a new level of understanding of environment–disease interactions. Public health needs would be served by use of selected species (including microbes) and habitats in early warning systems through collection, validation, and dissemination of environmental measurements and integration of earth observations, oceanographic, epidemiologic, and socioeconomic data. However, such trans-disciplinary data collection and analyses are not currently taking place; indeed, much of the data needed for such an approach do not exist. Issues of particular concern, which require multidisciplinary data to facilitate public health protection, include microbial and chemical contamination of coastal waters and fisheries; impacts of HABs on coastal communities; and effects of climate change on human health, storm frequency and intensity, and the prevalence of HABs and pathogenic microorganisms.

To protect health and enable epidemiological analyses in the future, the Federal Government should prioritize making the collection of human and animal health surveillance data integral parts of ocean observing.

Following recommendations of the international Group on Earth Observations, Federal agencies need to collaborate with the World Meteorological Organization and other international bodies to explore how observing systems may respond to public health community needs by collecting, providing, and validating remotely sensed environmental measurements relevant to human health.

Research Priority 2: Understand Human Health Risks Associated with the Ocean and the Potential Benefits of Ocean Resources to Human Health

While many ocean-based benefits and risks to human health are known, most are poorly understood and their direct health effects on humans are inadequately documented.42 Benefits to humans associated with consumption of seafood (e.g., improved cardiovascular health, cognition) should be better characterized.

Investigations should also quantify risks and impacts of exposure to health hazards (e.g., contaminants, pathogens, toxins) via various pathways, including skin contact, respiration (inhalation of airborne materials), and consumption of food or water; and determine the incidence and severity of associated human illnesses. For example, understanding the trade-offs between the health benefits of seafood consumption and the risks of mercury exposure to certain population groups would be valuable to consumers.

Efforts to quantify risks and benefits require improving assessment methods as well as the accuracy and sophistication of epidemiological studies. New epidemiological studies focusing on risks should focus on the most vulnerable human populations (children and pregnant women), individuals with compromised respiratory or immune systems, people who spend a large amount of time on or in the ocean, coasts and Great Lakes (e.g., commercial, recreational, and subsistence fishers), and on diseases in species that may serve as sentinels for new or ongoing threats. Ocean data and modeling should be integrated with assessments and epidemiological studies to define exposures and focus risk assessments.

Research Priority 3: Understand How Human Use and Valuation of Ocean Resources Can Be Affected by Ocean-Borne Human Health Threats and How Human Activities Can Influence These Threats

Economic and social studies are needed on how humans use and value the health benefits provided by the ocean. Coastal, ocean, and Great Lakes environments are highly attractive places for recreation and they play major roles in the Nation's diverse culture and economic vitality. However, the use and value of these environments and their resources may be significantly impacted by water-borne threats. These impacts include, but are not limited to, beach closures, constraints on resource use (e.g., fishery and shellfish harvest area closures), reduced marine operations and recreation due to contamination by pathogens, toxins, or pollutants, and other human health threats (real and perceived). Human activities, such as intensive agricultural operations and coastal development, contribute to the onset and persistence of many of health threats. Understanding and predicting the relationship between social and economic drivers (e.g., shoreline development, tourism, recreational and subsistence fishing) and threats to human health and quality of life will require integration of socioeconomic data and investigations with ecosystem-based studies of health threats at appropriate temporal and spatial scales. This integration will, in turn, help support management and mitigation efforts. There is a need for more social, behavioral, and economic research that encompasses both positive and negative economic and socio-cultural outcomes related to ocean health risks and benefits.

Research Priority 4: Apply Understanding of Ocean Ecosystems and Biodiversity to Develop Products and Biological Models to Improve Human Well-Being

The intricacies and diversity of ocean ecosystems provide unparalleled opportunities for discovery and development of useful products and technologies. The discovery and development of new pharmaceutical and other beneficial products have great potential. Collaborative research efforts that incorporate multiple disciplines (e.g., evolution, ecology, pharmacology) should focus on expanded assessments (e.g., functional genomics, advanced non-culture-based methods), and development (e.g., biosynthesis) of ocean bioproducts (e.g., pharmaceuticals, nutrients, diagnostic tools, reagents, enzymes). Research efforts should also include developing the capability to use marine species as models for the study of diseases, toxicology, and biochemical processes relevant to human health; and identifying and using appropriate sentinel species (e.g., aquatic, avian, mammalian) and habitats that may serve to give early warning of potential ocean risks to humans.

Tools to Be Emphasized

A comprehensive HEWS should be prioritized to integrate near-real-time physical, chemical, and biological data with records of illness due to pathogens, biotoxins, and chemical pollutants. For example, monitoring marine mammal health and mortality patterns is an excellent way to screen for potential human disease vectors. Integral to this system is an improved program of environmental surveillance that is well-coordinated with ocean observing infrastructure through the Integrated Ocean Observing System. Such a warning system would encompass collected human, habitat, and animal health data, as well as relevant environmental data (sea surface temperature, salinity, chlorophyll, nutrients, pH, aragonite saturation) via remote, satellite, moored, and mobile platforms and sensors as part of an integrated ocean-observing system. These efforts should be coupled with new data integration and interpretation capabilities and rapid data-communication plans. Once in place, the system could monitor, forecast, and respond quickly, issuing public advisories when ocean conditions occur that place people, domestic animals, and wildlife at risk from pollutants, contaminants, HABs, or disease. We should improve or develop sensors and diagnostic tools that are capable of rapidly detecting and identifying biological and chemical parameters, including microbe and pathogen densities, biotoxins, and contaminant concentrations. Remote sensing of ecological changes and real-time, high-frequency (temporal and spatial) *in situ* monitoring combined with statistical-empirical and mechanistic modeling are necessary to inform understanding of ecosystems and to track and predict outbreaks and other impacts. We need improved coordination of toxic algal-bloom monitoring, pathogen-source tracking, marine-disease surveillance, and illness reporting. A national suite of contaminant standards (e.g. polycyclic aromatic hydrocarbons, dispersant products) and standards for monitoring and screening should be cooperatively developed and implemented by Federal agencies, States, and regional ocean alliances.

Development and incorporation of sensors for pathogens, indicator microorganisms, harmful algae, toxins and other contaminants, and water quality into observing systems at local to regional scales will allow acquisition of data essential to support public health applications and decision making. There should be a focus on collecting data in nearshore environments, where the majority of the public's interaction with coastal waters occurs, and on recording environmental observations at appropriate temporal and spatial scales, particularly those relevant to the population dynamics of waterborne pathogens and harmful algae. In addition, statistical and bioinformatic43 tools need to be developed to link existing and new oceanographic models with less well-developed models of human health exposure and disease.

Designing advanced sensors and collecting devices for hostile and extreme (e.g., deep water) ocean environments will enable more extensive exploration of habitats and associated species, including bacteria, fungi, and viruses, and the discovery of potentially novel compounds and biochemical pathways. Developing, improving, and applying new methods and tools at shore-based marine laboratories and other facilities with specialized instrumentation (e.g., for large-scale gene sequencing) and computational resources in areas such as genomics, proteomics, and bioinformatics will expand survey and screening capabilities.

Tools such as microarrays, genetic bar-coding, and rapid DNA sequencing provide the means to identify previously unknown pathogenic species and identify incipient disease processes at cellular and molecular levels. Federal agencies should prioritize cyberinfrastructure that enables efficient data flow among stakeholders, information integration and synthesis, and complex modeling. We should continue to develop models that couple observations of environmental change with social and economic valuations, along with new visualization tools and decision-support techniques that effectively communicate complex information in ways that the public and policy makers can understand and use. These new technologies should be transferred into social management programs that advise the public and protect public health and the health of ocean, coastal, coastal watershed, and Great Lakes ecosystems.

Regional Applicability

The coastlines and offshore areas of our oceans and Great Lakes vary greatly in their physical, chemical, and ecological attributes and their living and non-living resources. States have recognized the need to cooperatively study and manage their shared coastal assets and watersheds and address common social, economic, and environmental issues that cross political boundaries. This recognition has prompted the formation of regional ocean alliances that now include coalitions among states that border the Great Lakes, northeast, mid-Atlantic, southeast, Gulf of Mexico, and west coasts. A similar strategy of shared regional environmental issues is driving the creation of Landscape Conservation Cooperatives (LCCs) among Federal agencies and

States. These cooperatives are envisioned as a network of self-directed partnerships that will inform resource management decisions relating to broad-scale environmental stressors, including climate change, within an overarching framework of adaptive management.

The formation of regional alliances and LCCs is evidence that the necessary science and management needs of our coasts are not "one size fits all." Each region has a unique suite of natural, social, cultural, and economic characteristics and priorities that influence the health of ocean, coastal, coastal watershed, and Great Lakes fish and wildlife resources, and ultimately the well-being of their citizens who depend on those resources. For example, temperature change in the Arctic is expected to be large relative to other regions.

Shorelines are shifting, homes are being undermined, and the distribution and abundances of marine resources are changing. Significant impacts are expected on subsistence fishing and the livelihoods of Native Alaskans. In the Gulf of Mexico, nutrient dynamics, HABs, shellfish contamination, and hypoxia may change significantly—for better or for worse—under different future climate scenarios.

In the southeast, the potential demise of coral reef ecosystems caused by high temperatures and ocean acidification may have detrimental impacts to recreation and tourism in Florida and the Caribbean. Environmental, social, and cultural changes are occurring and affecting human health and the economic well-being variably by region.

CROSS-CUTTING TOPICS

In this report, *Science for an Ocean Nation*, the list of recommended tools and mechanisms for change outlined in *Charting the Course* has been expanded into a larger set of cross-cutting topics. These are so named because they cut across the societal themes and have significant bearing on all of the research priorities.

They are discussed here in terms of Foundational Issues, Elements of Capacity, and Collaborative Approaches. Attention to the cross-cutting topics will be essential for addressing the research priorities in the societal themes and the priority objectives in the National Ocean Policy.

Foundational Issues

Global Climate Change

Central to *Science for an Ocean Nation* is the recognition that climate change is having a pervasive impact on ocean, coastal, coastal watershed, and Great Lakes ecosystems. Climate change stressors are likely to make ecosystems more vulnerable to the effects of human activities, which together could create abrupt changes that are disruptive to coastal communities and economies. The first foundational issue is that developing the observations, insight, and modeling capability needed to forecast the effects of climate change on ecosystems and the services they provide is essential to both sustained use and wise stewardship.

Overview of Status

Earth's life-support systems are threatened by global climate change. The effects of climate change, including increasing water temperature, rising sea level, altered hydrologic patterns, loss of sea ice, and ocean acidification, are expected to further degrade vital ecosystem services, putting at risk lives and livelihoods. Scientists are challenged to reduce uncertainty about the causes and impacts of environmental change, whether natural or anthropogenic, and to communicate such knowledge broadly in effective ways. The strength of our Nation depends on a robust economy, sustainable development, human health, national and homeland security, and energy independence. Each of these priorities may be affected by the potentially destabilizing impacts of global climate change. Climate change science is inherently multidisciplinary and crosscutting. It is inspired by the recognition that the Earth's ecosystems and the services they provide are complex, interdependent, and increasingly dominated and threatened by human activity. Our influence on the mobilization and flux of carbon, as well as nitrogen and other elements within the biosphere, impacts the capacity of natural recycling processes, resulting in, for instance, dead zones in the Gulf of Mexico and Chesapeake Bay. Changes in air and water temperatures, sea levels, precipitation patterns, ocean chemistry, storm intensity, flooding, and ice melting are expected to continue impacting our water resources, food production, fish and wildlife habitats, endangered species, homes, industries, recreation, essential infrastructure, and public health and safety.[44] We are making progress in addressing some of these complex issues. We are beginning to comprehend the full nature and extent of climate change. We are refining our understanding of how water, carbon, and nitrogen cycles and ocean productivity are influenced by climate. Strategies to reduce and mitigate the accumulation of atmospheric CO_2 using alternative energy sources, energy conservation, and carbon sequestration in biological and geological reservoirs are being studied intensively. There is increasing public recognition that Earth's ecosystems are interconnected and interdependent.

People understand that choices made about such things as cutting forests, conserving energy, using sustainable energy sources, recycling, and reducing carbon emissions, all play a role. The ocean research and resource management communities are speaking with one mind about how science should inform management and how decision-support tools should encompass the explicit objectives of multiple stakeholders.

There is agreement over the need for integrated, multidisciplinary observations, monitoring, research, and downscaled modeling, along with effective assessment, synthesis, translation, and communication of information among all stakeholders. Improving science and ocean literacy is a critical overarching goal.

Ultimately, the stewardship of our freshwater and marine environments and the persistence of biologically diverse, healthy ecosystems that support healthy people and a vibrant economy will depend on how we collectively view and respond to a world of changing environmental, social, and economic knowledge, values, and priorities.

Supporting National Priorities

Box 6: Supporting the National Ocean Policy: Climate Change

A central message of the National Ocean Policy is better coordination among institutions at all levels to improve understanding and inform decisions about the open ocean, coasts, coastal watersheds, and Great Lakes. This will allow effective response to the challenges of mitigating and adapting to global climate change and ocean acidification with comprehensive marine planning and ecosystem-based management. Stewardship of our resources will depend on strengthened and integrated observing systems, effective information dissemination, new tools to model and forecast change, and policies that improve land use and water quality and that build resiliency, enabling the wise use, protection, and sustainability of our resources.

Beach near the port of Galilee, Point Judith, Rhode Islands. Credit: Mr. William B. Folsom, National Marine Fisheries Service, National Oceanic and Atmospheric Administration.

Administration priorities include sustainable development, economic growth, energy independence, human health, and national and homeland security. Achievement of these goals requires the recognition that our livelihoods and wellbeing depend on the vital goods and services provided by natural ecosystems; that human activities are changing climate by altering the characteristics of the ocean, air, and land and their interactions, thereby degrading ecosystem life-support functions; and that changes in human attitudes and behaviors should go hand-in-hand with informed and effective management of our environmental, social, and economic resources (see Box 6).

Making Progress

While we have made important strides in understanding how, why, and in what ways the ocean, our coasts and coastal watersheds, and the Great Lakes are changing, we still lack fundamental knowledge about how climate factors alone, and in interaction with other stressors, will impact the viability of ecosystem functions and services. Many questions remain. How much of our coastal landscapes and seascapes should be maintained in a natural state to sustain essential ecosystem services? How will climate change affect biodiversity and ecosystem function and resilience? What priority management actions are necessary at local, regional, and global levels to conserve ocean, coastal, coastal watershed, and Great Lakes resources; protect shoreline infrastructure; and sustain our health, safety, security, and wellbeing? How will increasing human populations respond proactively and adapt to impending, irreversible climate change? How will ocean acidification impact the structure and function of marine ecosystems? What are the current ecosystem thresholds for irreversible change?

Climate change is pervasive, cumulative, multi-dimensional, and difficult to measure and quantify. The physical, chemical, and biological drivers of ecosystem dynamics are complex; various environmental stress-ors may interact in poorly known, non-linear ways. To continue making progress Federal agencies should:

- Establish reliable baselines for analyzing variability in many system states (differentiating between natural and anthropogenic change is often problematic);
- Describe global trends and inform managers about changes that will occur at local levels, for example, relative sea-level rise that varies regionally with geological uplift/subsidence, winds, currents, temperature, salinity, and in wetlands, plant growth;
- Improve decision-support tools to enable decision makers to envision future environmental, social, and economic conditions at global to local scales, and allow them to test various management options and priorities;
- Provide decision makers with analyses of impacts based on the best available scientific information, including a range of likely conditions. Multiple stakeholders will need to be involved early in defining needs and in the information gathering and analysis process; and
- Engage and inform the public. Human behaviors and attitudes are not easily understood or changed. Regardless of our level of scientific understanding, our ability to effectively mitigate, build resilience, and adapt to climate change, while minimizing social disruption, will depend on public engagement.

Social Sciences

Adaptation to, and mitigation of, the coupled effects of climate change and human use on ocean, coastal, coastal watershed, and Great Lakes environments will require far more than the integration of physical and biological sciences. Making the difficult trade-off decisions that are certain to come will require entirely new ways of valuing ecosystem services, understanding decision-making with respect to risks, and fostering individual and collective change. These are issues that can be usefully addressed by the social sciences, including economics, sociology, psychology, and anthropology. Therefore a second foundational issue is realizing effective solutions to environmental challenges by seamlessly integrating social sciences with the natural sciences, and developing a larger workforce committed to working at this interface.

Overview of Status

Economic prosperity, human health, and societal well-being are intrinsically tied to the health of our ocean, coastal, coastal watershed, and Great Lakes ecosystems. The National Ocean Policy identifies eight major categories of commercial use of these resources: oil and gas, renewable energy, mining, ports and harbors, commerce and transportation, tourism, aquaculture, and fishing. This report also notes that these ecosystems provide a range of other services including scientific, recreational, cultural, conservation, and homeland national security activities; and are home to unique national trust resources valued by the public, such as protected species, wetlands, and corals.

Social, behavioral, and economic sciences help us to understand ecosystem use patterns as well as societal preferences on the provision of ecosystem services. Collectively, these disciplines can inform us about the magnitude of the benefits derived from ocean, coastal and coastal watershed, and Great Lakes ecosystems, clarify the costs and benefits of proposed management actions; and assist in forecasting future use patterns, including responses to management measures and changes in environmental conditions. Major Federal mandates that cover ocean, coastal, and Great Lakes resources include but are not limited to the National Environmental Protection Act, Executive Orders 12866 (Regulatory Planning and Review) and 12898 (Environmental Justice), the Endangered Species Act, the Regulatory Flexibility Act, the Clean Water Act, and the Marine Mammal Protection Act. To comply with these Acts, a suite of socioeconomic assessments is required that includes cost-benefit analysis, social impact assessment, cumulative impact assessment, small business impacts, and social justice assessment.

The principal threats to our ocean, coastal, coastal watershed, and Great Lakes ecosystems stem from anthropogenic causes; hence any action taken to protect or restore ecosystem health should consider human drivers as well as environmental information. Social sciences help us understand the drivers of human action and evaluate societal tradeoffs and predict responses to change in policies, environmental conditions, and incentives (e.g., changes in prices or income).

A wide range of methods and models is used by social scientists. At a minimum, we need social and economic information on ecosystem use patterns. This information suite ranges from basic economic performance measures, such as profits, productivity, and value added, to gross domestic product, to improved social indicators of well-being including vulnerability and resilience metrics.

Policy makers, however, require more than descriptive information on status and trends of ecosystem use.[45] To ensure sustainable policies and use of our ocean, coastal, coastal watershed, and Great Lakes ecosystems, the social sciences should provide information on societal values of ecosystem services, the impact of human uses on ecosystem function, and the social and economic trade-offs and time horizons of benefits associated with alternative use patterns. Emerging methods such as integrated ecosystem assessments provide a rich characterization of the linkages and feedback mechanisms among a broad range of ecosystem components, based on models of biological, ecological, and oceanographic processes coupled with economic models of human use patterns and preferences.[46] In addition, economic models of spatial choices are increasingly incorporating biological and ecological information, including water quality. These models can be used to assess the value associated with activities in an area; estimate changes in values due to perturbations of environmental factors or changes in allowed uses in that area; and predict how users will respond to environmental or management changes. Importantly, the models can also incorporate risk preferences and environmental uncertainty, a defining ecosystem feature.

Valuation surveys are useful tools for valuing goods that are not traded, such as threatened and endangered species, cultural resources, and habitats, such as corals and wetlands. While these methodologies are not new and have a lengthy history in the peer-reviewed literature, they have not been extensively used. Consequently, relatively few resources have actually been valued. In a regulatory setting, this often means that while the cost to industry of recovering an endangered species can be estimated and provided to decision makers, a value has not been set on the benefits to society from rebuilding stocks. More recently, public and stakeholder surveys have been used to determine societal preferences for proposed management actions. Often referred to as "conjoint surveys," these have been used in ocean and coastal settings to assess angler preferences for management alternatives as well as to assess public preferences for allowed uses in marine managed areas.

Increasingly, sound social, behavioral, and economic science-based approaches are needed to understand how households, communities, and regions can become more resilient under the threat of natural hazards. An understanding of how decision makers respond to uncertain information should inform the kinds of information products to be produced. Additionally, knowledge of the community and economic drivers that fuel resilience and recovery after a destructive event should be used to direct resources to further strengthen an affected region. Quantifiable measures are needed to gauge progress. Resilience and vulnerability indices of infrastructure, community, and economic fitness can be used as performance measures. Regional economic models serve to identify the local engines of economic recovery. Survey techniques can be used to value important cultural and recreational amenities and to develop a better understanding of individual risk perceptions that lead to adaptation actions.

Box 7: Supporting the National Ocean Policy: Social Sciences

Strong social science capabilities are necessary to implement the National Ocean Policy. A foundational principle of the Policy is that current and future use of ocean, coastal, and Great Lakes ecosystems should be managed and effectively balanced (i.e., considering trade-offs) in a way that maintains and enhances the environmental sustainability of multiple uses, including those that contribute to the economy, commerce, recreation, security, and human health. Social science research is uniquely capable of identifying management options that provide the greatest societal benefits while achieving conservation goals.

Beach near the port of Galilee, Point Judith, Rhode Islands. Credit: Mr. William B. Folsom, National Marine Fisheries Service, National Oceanic and Atmospheric Administration.

Supporting National Priorities

The National Ocean Policy provides a comprehensive and integrated framework for achieving sustainable use of our ocean, coastal, coastal watershed, and Great Lakes ecosystems. Marine planning cannot be effectively implemented without substantial input from the social sciences. Marine planning requires cost–benefit analyses of alternative uses in spatially-defined areas; the ability to predict how affected users will respond to changes in allowable uses within an area; and public and stakeholder buy-in. Social science methods include a variety of models for conducting cost–benefit analysis and predicting user responses to spatial management strategies. Stakeholder and public preference surveys will strengthen buy-in from these groups by providing a transparent process for incorporating multiple objectives and multiple values into decisions. As these techniques encourage and improve stakeholder input, marine planning processes will incorporate stakeholder values, beliefs, and concerns that it would not be aware of otherwise. These tools will also identify and help design necessary social products and services and improve prioritization of projects.

Through integrated modeling, social sciences can directly contribute to ecosystem-based management. Integrated modeling that incorporates social sciences can also identify cost-effective ecosystem restoration programs, assess economic and social effects of climate-induced changes in the Arctic, and assess and predict the economic consequences of proposed land use strategies to the benefits derived from our ocean, coastal, coastal watershed, and Great Lakes ecosystems.

Incorporation of social science analysis in these efforts also will support the broader national priorities (see Box 7). The application of social sciences can support sustainable development in coastal and Great Lakes regions, thus contributing directly to the overall economy and human health. Marine planning-related economic research will identify the costs and benefits associated with traditional and renewable energy use patterns in these regions, providing crucial information for the Nation's energy security. Finally, climate-related research can assess the socioeconomic impacts of climate change as well as the cost-effectiveness of adaptation activities necessitated by climate change.

Making Progress

The significant and increasingly competitive uses of our ocean, coastal, coastal watershed, and Great Lakes ecosystems suggests a level of competition for space and resources not previously considered by any single Federal agency. As demonstrated herein, social science can play a critical role in determining how best to manage and utilize our resources. To provide this information, however, requires a stronger social science capability underpinned by socioeconomic data. Because of rapid changes in the Arctic, information on communities in this region and interactions with the natural environment is especially critical. Requirements include:

- Economic and social information on individuals, households, communities, and coastal and marine-related sectors for tracking performance, understanding current use patterns, and predicting responses to management measures and changes in environmental conditions;
- Capability to conduct integrated ecosystem assessments, including providing richer characterization of socioeconomic interactions and feedback mechanisms;
- Predictive models for estimating changes in ecosystem services values and trade-offs based upon standard climate scenarios;
- A modeling toolbox that would allow managers to evaluate changes in benefits and the distribution of benefits from proposed marine planning, for example multi-sector location choice models;
- Surveys of public preferences regarding coastal and marine management strategies;
- Habitat valuation surveys to determine societal preferences for habitat uses and conditions;
- Protected species valuation surveys to assess societal values and preferences for alternative recovery programs;
- Social capital surveys to identify social resilience networks that improve productivity and contribute to positive economic outcomes;
- Identification and assessment of community risk factors, assessment of impacts of management options on risk, and identification of cost-effectiveness of strategies for alleviating risk; and
- Development of community resilience indicators.

Ocean Literacy

The third foundational issue emphasizes development of a multi-disciplinary ocean workforce and an ocean-literate citizenry that better understand the majesty and mystery of the ocean, our coasts, and the Great Lakes, and make well-informed decisions about their use. Ocean educators are working at all levels of formal education and in informal venues such as aquaria, museums, and in the media. Rapid improvements in data-streaming, visualization, and cyber-enabled communications should be used for education as well as research. Improving education and ocean literacy will involve a long-term commitment to change, beyond tomorrow's ocean scientists, managers, and decision makers. Public perception shifts with familiarity; people protect what they understand and perceive as valuable. Familiarity with healthy ecosystems and the tangible and intrinsic values provided by them are no exception. Education and understanding among the ocean workforce and the general public regarding the importance of ecosystem services that benefit both society and individuals will, over time, alleviate anthropogenic pressures and concomitant ecosystem degradation.

Overview of Status

Federal science and resource management agencies invest in students, educators, and the public by supporting science, technology, engineering, and mathematics (STEM) education as well as environmental education. Investments capitalize on agency assets, such as the outdoor classrooms provided by national parks, estuarine reserves, and marine sanctuaries, and the technological capabilities of ships, laboratories, and ocean observing platforms. Ocean education programs contribute to the President's Educate to Innovate Campaign and the Strategy for American Innovation by connecting students and teachers to world-class research and emerging technology that are the foundation of our future economy. Federal ocean education programs include teacher professional development; partnerships with aquaria and museums; student scholarships, fellowships, and internships; partnerships with minority-serving institutions; and direct contact with audiences at parks, sanctuaries, and reserves. For most agencies, ocean education is part of the larger portfolio of STEM, earth systems science, and environmental education.

A recent public survey by the Ocean Project showed that Americans continue to support ocean conservation, but rank it as a low priority and do not understand issues or underlying science concepts. Young people (aged 12–17) know and care more about ocean and other environmental issues, and they are more willing to act than adults. Furthermore, they influence the opinions of adults, who tend to view their children as better informed on conservation issues.[47] This survey provides a baseline against which we can measure our impact on future public understanding of ocean issues.

In 2004, a coalition of educators and scientists joined together to develop "Ocean Literacy: The Essential Principles of Ocean Sciences, K-12, "[48] a defining framework for formal and informal educators. This framework identifies seven essential principles that ocean-literate citizens should understand. In 2009, a companion piece, "Climate Literacy: The Essential Principles of Climate Sciences," was developed and adopted by the U.S. Global Change Research Program. Over the past several years, Federal agencies have successfully expanded the use of these materials across a partnership network that includes the Smithsonian's Sant Ocean Hall, the Centers for Ocean Science Education Excellence, the Coastal Ecosystem Learning Center aquaria, and the Science-On-a-Sphere museum network, in addition to many other Federal and non-Federal partners.

Federal agencies also oversee a variety of programs to build our future ocean workforce. Students are supported through internship, fellowships, and scholarship programs. Federal agencies collectively support the National Ocean Sciences Bowl, which has grown to include 25 regional teams and 300 schools, involving over 2,000 high school students each year. Programs at minority-serving institutions provide direct student support, mentoring, and career development opportunities. Agencies also fund studies to track the ocean-related workforce over time. Support for ocean education today will shape the scientific and technical workforce of the future. The Regional Associations of the Integrated Ocean Observing System report that 32 percent of their employees have a bachelor's degree, 26 percent have a master's, and 26 percent have a PhD.[49] Although the numbers of PhDs awarded in the geosciences has remained steady, the numbers of Bachelor's and Master's degrees have fallen since the 1980s. One area of success has been the steady increase in the number of women obtaining degrees in geosciences, with approximately 43 percent of degrees obtained by women in 2008. Efforts to increase minority participation are still needed, as African Americans and Hispanics continue to make up less than 10 percent of the geosciences workforce.[50]

Supporting National Priorities

Ocean education is closely linked to the President's efforts on energy independence, sustainable development, climate, health, national and homeland security, and the economy. All of these efforts depend on an educated, diverse, multi-disciplinary workforce that understands the ocean and its impact on our society (see Box 8). This workforce should reflect the U.S. population in terms of diversity of individuals, backgrounds, cultures, and perspectives. The education efforts of each agency contribute to the President's Educate to Innovate campaign and the President's Strategy for American Innovation.

Box 8: Supporting the National Ocean Policy: Ocean Literacy

The National Ocean Policy includes fostering a public understanding of the value of the ocean, coasts, and Great Lakes to build a foundation for improved stewardship. Promoting and improving ocean literacy is critical to public engagement in and understanding of ecosystem-based management, marine planning, and climate impacts and mitigation, including in the Arctic. Increasing knowledge among the public, decision makers, and stakeholder groups builds awareness of connections between human activities on land and in the water and resource health and availability. Undergraduate and graduate education of future scientists ensures continued progress in understanding ocean, coastal, and Great Lakes environments and our interaction with them. Use of ocean infrastructure for educational purposes helps promote public and K-12 understanding.

Beach near the port of Galilee, Point Judith, Rhode Islands. Credit: Mr. William B. Folsom, National Marine Fisheries Service, National Oceanic and Atmospheric Administration.

The mission science agencies are committed to support the Educate to Innovate campaign, a nationwide effort to move American students from the middle to the top of the pack in science, technology, engineering, and math over the next decade. The President identified three overarching priorities for STEM education: increasing STEM literacy and enabling critical thinking; improving the quality of teaching in these fields so American students are no longer outperformed by those in other nations; and improving STEM education and career opportunities for underrepresented groups, including women and minorities. The Innovation Strategy identifies education as a cornerstone for future innovation and competitiveness. The strategy proposes to invest in STEM education at all levels. Ocean education programs contribute by linking students and teachers to exciting discoveries and innovative technologies.

Making Progress

To continue progress on this foundational issue, we should:

Continue to focus on educating students, educators, and the public on the essential principles of ocean literacy including stewardship, natural hazards, the ocean's role in climate, ocean ecosystem health, and the ocean's role in human health. These programs work best on a state and regional basis, as the educational standards of each state and the ecosystems and natural hazards of each coastal region are different;

- Encourage and support students to obtain degrees in oceanography or related sciences, particularly minority students who wish to enter these fields;
- Develop pipeline programs that reach students at community colleges and that bridge transition points between high school and associate degree programs to college, and graduate schools. Mentoring programs are necessary to increase enrollment of all students in geosciences; and
- Conduct baseline studies of environmental literacy and investments in evaluation methods to adequately assess the effectiveness of different types of programs.

Elements of Capacity

Ocean Observations and Infrastructure

Ocean conditions are changing over the long term and being affected in the short term by human-caused and natural events. Effective adaptation and management require ocean observations on decision-relevant time and space scales, and the infrastructure to support those observations. Together, infrastructure and observations enable baseline knowledge of areas and processes ranging from microscopic to global, record changes, and document the impact of management and mitigation practices. Impressive new capabilities are emerging for making *in situ* observations from fixed and mobile ocean platforms. At the same time, consideration should be given to maintaining vessel and satellite capability. Both science and policy call for improving integrated ocean observing capability, across the United States and internationally.

Overview of Status

Steady changes such as rising sea level and ocean acidification, together with episodic events such as the Deepwater BP oil spill, are having profound effects on ecosystems and society. The understanding needed to manage, mitigate, and adapt to change requires a wide variety of coordinated observations and analysis using a broad range of tools. These encompass remote sensors mounted on satellites and shore facilities; *in situ* ocean-observing systems; ships and autonomous vehicles; and computers, data storage, and service to provide rapid data access. Advances already underway in ship technology (e.g. acoustically quiet vessels to monitor fish stocks), tracking devices, and *in situ* environmental monitoring systems allow for increased observation capacity. Increasingly there is movement toward automated or remote systems, which greatly augment the scope and breadth of information collected. Real-time tracking of environmental conditions, especially with the aid of satellites, facilitates rapid and targeted response to emerging situations, both acute, such as natural hazards, and chronic, as with climate change. Beyond making environmental data collection and observation possible, these tools also provide the foundation for resource use and management, transportation, commerce, security, and surveillance.

Progress on the research priorities associated with each one of the six societal themes outlined in *Science for an Ocean Nation* depends on ocean infrastructure and observation. Existing infrastructure is already being leveraged to support these priorities, but as the recent oil spill in the Gulf of Mexico has shown, technological advances are continually required to meet the needs of researchers, managers, regulators, and policy-makers. In the National Ocean Policy, one of the five areas of special emphasis is mapping and infrastructure, which are needed to meet all other national priority objectives.

It is critical that Federal agencies plan cooperatively to ensure that, on both short and long timescales, the United States has sustained adaptive ability to gather, analyze, and deliver ocean data. Since *Charting the Course* was published, Federal legislation, a new National Ocean Policy, several new development or construction projects, and several coordinated planning activities are clarifying priorities. The Omnibus Public Land Management Act of 2009 (Public Law 111-11) directs various Federal agencies to establish a coordinated national ocean exploration program; develop a coordinated and comprehensive Federal ocean and coastal mapping plan; establish a national system of ocean, coastal, coastal watershed, and Great Lakes observing systems; develop a strategic research and monitoring plan to guide Federal research on ocean acidification; and create a coastal and estuarine land conservation program for protecting important coastal and estuarine areas.

Some noteworthy new ocean infrastructure and observing capability is coming on line or is under development. The global-class research vessel *R/V Okeanos Explorer* became available in 2009, the ice-strengthened *R/V Sikuliaq* is under construction and is expected to begin science operations in the Arctic in 2014, and the Navy has announced operators for two new ocean-class ships also expected to begin operations in 2014. A new acoustically quiet fishing survey vessel, the *FSV Shimada*, began surveys on the West Coast in 2010 and the *FSV Reuben Lasker* is under construction.

Ships are now designed to be multi-use platforms able to conduct fisheries research and bathymetric surveys, and collect oceanographic data simultaneously. It is imperative that these platforms be able to “collect once, use many times” and share the data. New sensors for ocean observations using fixed or mobile platforms are in development, prototype, or commercial stages, partly as a result of the Near-term Priority for Ecosystem Sensors identified in *Charting the Course*. Working with Federal agency and regional partners, NOAA is establishing a robust program for the Integrated Ocean Observing System to integrate existing regional observing capabilities, while construction has started on the National Science Foundation’s (NSF) Ocean Observatories Initiative, designed to address basic ocean research questions. Both are working toward coordinated data management to provide transparent information access and delivery. On the satellite side, new capabilities (e.g., for imaging sea-surface salinity) will be launched in the next several years. However, consideration should be given to maintaining satellite sensor capabilities and global ocean and coastal datasets (e.g., of ocean color).

Recent planning and assessments will provide important input to the strategic action plan for observation, mapping, and infrastructure called for in the National Ocean Policy. An inventory by the SOST Interagency Working Group for Facilities in 2009 found that current ocean infrastructure was adequate to address the Near-term Priorities identified in *Charting the Course*, but lacked future capacity. Also in 2009, the National Research Council

(NRC) published *Science at Sea: Meeting Future Oceanographic Goals with a Robust Academic Research Fleet*, calling for improved use of autonomous vehicles and further development of global and regional-class research vessels, with a judicious mix of general and special-purpose designs. A task force on unmanned systems has been created to focus on needs with respect to unmanned platforms, which are becoming an important tool for oceanographic data collection. A SOST-funded NRC study provides advice on the types of infrastructure needed to facilitate ocean science and research in 2030.[51]

Supporting National Priorities

To make informed and effective decisions regarding economic, development, and security activities, human and ecosystem health and welfare, and mitigating and adapting to a changing environment, decision makers and the public should have ready access to tools and information based on data gathered through ocean observations and infrastructure (see Box 9). This is especially critical as climate change brings about environmental change, globally and with particular impact in sensitive regions. The changing conditions in the Arctic have prompted increased efforts in the development of Arctic-class vessels that can safely and responsibly operate in the newly opened waters. The risks associated with offshore oil and gas exploration and extraction from deeper waters and new offshore areas, and development of renewable energy sources from the ocean, make precise real-time ocean observations critical for safe and efficient energy operations. Data are required to support a wide range of both commercial and safety decisions, including evacuations, beach and shellfish closures, establishing science-based annual catch limits for fisheries, and the identification of safe and efficient shipping routes to transport the products we rely on each day. An integrated observing system including satellites, research ships, and autonomous platforms will link observations to modeling and prediction, enabling development of the information products needed to:

- Improve predictions of climate change and weather;
- Improve the safety and efficiency of maritime operations;
- Allow more effective mitigation of the effects of natural hazards;
- Improve national and homeland security;
- Reduce public health risks;
- Allow more effective protection and restoration of healthy coastal ecosystems; and
- Enable the sustained use of ocean and coastal resources.

The oceanographic fleet also responds to crisis events to help ensure human and environmental health, as seen in the response to recent natural hazards and environmental tragedies.

Box 9: Supporting the National Ocean Policy: Observations and Infrastructure

A central element of the National Ocean Policy is infrastructure to provide the data and tools necessary to support responsible management of the open ocean, coasts, coastal watersheds, and Great Lakes. The strategic action plan to be developed for this priority objective will address a nationally integrated system of observing systems, the use of unmanned vehicles, remote sensing platforms, and satellites, and the identification of capabilities and gaps of the national oceanographic fleet of ships and related facilities.

Beach near the port of Galilee, Point Judith, Rhode Islands. Credit: Mr. William B. Folsom, National Marine Fisheries Service, National Oceanic and Atmospheric Administration.

Making Progress

The Nation's facilities and infrastructure are the foundation for accessing, obtaining, analyzing, and sharing information needed to support not only the six societal themes presented in this document but also resource management decisions and policy-making. Central to making progress are the needs to integrate multiple types of observing capabilities into a comprehensive system that can rapidly deliver critical data, to ensure records of change in the form of high-quality time series, and to leverage U.S. and international investment to provide global coverage.

To accomplish the goals of Science for an Ocean Nation and the proposed ocean policy priorities, progress should be made in the following areas:

- Integrate and improve existing data-delivery systems and modeling capabilities at regional, national, and global scales. Data should be rapidly and freely accessible through a comprehensive national data network, providing transparent metadata and analysis tools for decision support and for research;
- Ensure a robust, integrated system of global, coastal, and Great Lake observatories in critical areas for monitoring ocean health, tracking climate change effects, and conducting climate-related ocean research, as well as for hazard prediction. This includes continuation of efforts toward a more capable ARGO system of profiling floats, and maximizing the use of ships of opportunity for global observations. Sensor miniaturization and commercialization for use across platforms, together with sustaining infrastructure for adequate calibration and validation activities across such networks, are particular challenges;
- Maintain continuity of global satellite ocean measurements and high-resolution coastal measurements (including but not limited to ocean color radiometry, scatterometry, and altimetry); develop coordinated virtual satellite constellations involving domestic and foreign assets; and establish domestic and international calibration and data validation programs;
- Ensure essential research vessel capabilities, including the appropriate mix of general and special purpose vehicles;
- Develop cost-effective approaches to maintain and provide access to the rapidly expanding fleet of ever-more-capable autonomous vehicles, perhaps comparable to the University National Oceanographic Laboratory Systems approach for shared use of research vessels and coordinated operations and maintenance; and
- Maximize impact of existing infrastructure in Arctic and Antarctic regions through improved operational coordination across agencies, planning and programming of infrastructure and shore facilities, and development of new technologies, such as under-ice navigational

capabilities for floats, gliders, and autonomous vehicles for study of sea ice dynamics and ecology.

Ocean Modeling

Transforming data into knowledge is critical to understanding and managing our coasts, the Great Lakes, and the ocean. Modeling is an essential tool in this transformation process.

Federal agencies are developing new abilities to model both climate and ecosystems on the scales at which decisions are made and implemented, and to construct more realistic approaches that integrate the chemical, physical, and biological processes that regulate the ocean and Great Lakes. Improved modeling and forecasting capabilities also require rapid and universal data access as well as automated assimilation of quality-controlled data.

Overview of Status

The capability to analyze and predict the state of the ocean, including physical, chemical, biological and ecological components, remains a key long-term goal for the Nation.

This need spans space and time scales from local bays and estuaries out to the entire global ocean, and from short-term forecasts out to decadal, centennial, and even millennial climate predictions. Numerical models are essential tools that can be used to analyze and predict the state of the ocean and to guide our interaction with it. They can be used diagnostically to better understand the relationships between the physical, biogeochemical, and socioeconomic aspects of the ocean, or they can be used to forecast the future impacts of global environmental change or proposed changes in management practices.

Coupling numerical modeling with biogeochemical and ecosystem models is particularly important to address the needs highlighted in the six societal themes, such as understanding the implications of ocean acidification, understanding the resilience of our coasts to extreme forcing from climate change or episodic natural hazards, assessing ocean ecosystem heath, and supporting ecosystem-based management.

The numerical tools required for simulating and predicting the ocean environment undergo constant upgrades, and in recent years there have been many new developments and improvements in a variety of ocean models and associated prediction systems.

Improvements have involved the use of new methods for more accurately representing the physical ocean, allowing models to run at higher resolution, and making computer models more efficient. Other improvements have been made in the data assimilation techniques used to incorporate observations of the ocean into the models to improve predictions. Many of these data-assimilative prediction systems are now operating in coastal regions around the country.

On larger spatial scales, there are global ocean models that assimilate observations from the Global Ocean Observing System to predict ocean conditions around the world.

Supporting National Priorities

While numerical models are useful for exploring and understanding the relationships between different dynamic systems, they also are of particular utility in understanding the implications of policy and management decisions by serving as a tool to predict the potential ramifications on the ocean and ocean-related systems (see Box 10).

These predictive capabilities are essential for forecasting the response of ecosystems to various types of forcing, understanding marine ecosystem function, assessing elements of the ocean's role in climate change, and understanding what ocean-related variables we should monitor (or build new sensors to observe) in order to accurately assess the state of the physical ocean and the marine ecosystem.

The Federal agencies' ocean modeling capability is key to forecasting ocean-influenced processes and phenomena, providing scientific support for ecosystem-based management, and guiding the efficient development of an ocean-observing system and related infrastructure that will be vital for understanding the ocean in our changing environment.

Box 10: Supporting the National Ocean Policy: Predictive Modeling

The need for a robust ocean modeling capability to guide ocean policy and management decisions is demonstrated throughout the National Ocean Policy.

To carry out the stewardship responsibilities put forth in the National Ocean Policy, the impacts of these decisions should be understood. Predictive ocean models are important tools for estimating consequences.

A broad program of basic and applied disciplinary and interdisciplinary scientific research, mapping, monitoring, observation, and assessment, coupled with development of forecasts, models, and other decision-support tools, is required to build knowledge of ocean, coastal, and Great Lakes ecosystems and processes and ensure that management and policies are based on sound science.

Beach near the port of Galilee, Point Judith, Rhode Islands. Credit: Mr. William B. Folsom, National Marine Fisheries Service, National Oceanic and Atmospheric Administration.

Making Progress

To support the six societal themes and the associated research priorities, improvements will be needed in our modeling tools. Priorities are to:

- Integrate existing modeling capabilities into a flexible and comprehensive unified ocean-modeling software environment, building on the diverse existing community and developmental ocean models;
- Conduct systematic best-practice studies to develop guidance in the selection of optimal algorithms and techniques for particular modeling applications;
- Improve biogeochemical models to increase our understanding of ocean acidification issues;
- Improve ecosystem models to better understand complex ecosystem dynamics and forecast the effects of resource use, exploration, and development on ecosystems and individual components;
- Continue to develop high-resolution, global ocean model configurations for use in real-time, short-term forecasts out to days or

months and reanalysis of historical ocean states, and multi-century climate forecasts;

- Provide regional and local models of environmental changes to help determine how these changes impact resources;
- Develop techniques to translate the output of ocean and climate system models, including probabilistic forecasts and uncertainty estimates, into information formats appropriate for input into decision-support tools;
- Continue to develop current computational capacity to assimilate increasing amounts of data provided by ocean-observing assets and to support and process model advancements that incorporate increased process and mechanistic understanding;
- Develop computational capabilities to support new modeling capabilities and to build and deliver decision-support information to relevant operational communities; and
- Conduct process-oriented research to resolve critical functional relationships encoded into models.

Collaborative Approaches

Mechanisms for Collaboration

Because resources and processes are trans-jurisdictional, ecosystem-based management of the ocean, our coasts, and Great Lakes should engender and embrace mechanisms for collaboration across many scales of government, between the public and private sectors, and with academic and private research partners.

Since release of *Charting the Course*, coastal states have strengthened existing regional associations and formed new ones, and are developing plans for identifying ocean and coastal priorities and needs. At the same time, there is an increasing emphasis on integrating climate and ecosystem models on the regional scale, where many policy and management decisions are made and implemented.

In addition to strengthening partnerships across different levels of governments, *Science for an Ocean Nation* emphasizes the importance of working regionally, strengthening interactions between resource managers and scientists, and developing new mechanisms to promote long-term partnerships among industry, government, academia, and non-governmental entities.

Overview of Status

Currently, marine resources and use are managed by different authorities, under varying regulations and separate governance structures but nature does not obey such boundaries.

Important new tools for improving our interactions with the marine environment, notably marine planning and ecosystem-based management, require consideration of both ecosystem components and the interactions among them.

For all marine resources to be managed under the same analytical framework, agencies with jurisdiction over differing ecosystem components will need improved mechanisms for collaboration and cooperation. Cooperation across political boundaries, including international lines, will lead to a more effective management framework and better understanding of the global ocean.

Currently, collaboration among Federal agencies is encouraged under the governance structure of the National Ocean Council. This cabinet-level body, comprised of representatives from Federal agencies with interests or activities in the ocean, is responsible for updating and setting national ocean priorities and overseeing implementation.

This structure also includes two major subgroups that allow Federal agencies to coordinate efforts on science and technology and resource management: the Ocean Science and Technology Interagency Policy Committee[52] and Ocean Resources Management Interagency Policy Committee. A governance coordinating committee consisting of representatives from States, Tribes, and local governance structures provides input on inter-jurisdictional issues related to the National Ocean Policy. Other ocean-related interagency collaborations can also facilitate ocean research priorities (e.g., the Committee on the Marine Transportation System, U.S. Coral Reef Task Force, and Great Lakes Interagency Task Force).

Mechanisms for regional collaboration exist on various scales. Regional organizations, such as the Gulf of Mexico Alliance and the Northeast Regional Ocean Council, are based on the needs of regions. For example, the Northeast Regional Ocean Council consists of six States and six Federal agencies and aims to promote regional protection and balanced use of resources. The Great Lakes Regional Collaboration includes Federal, state, tribal, local, and international groups working together to design and implement a strategy for preservation and restoration of the Great Lakes. The West Coast Governor's Agreement on Ocean Health brings States and Federal agencies together to protect and manage marine resources along the entire West Coast. These regional mechanisms allow for all levels of governments to communicate and work together for common goals.

Several mechanisms currently exist that facilitate collaboration across multiple sectors. These mechanisms include formalized interagency activities designed to promote public–private research and education partnerships (e.g., the National Oceanographic Partnership Program) or to pool capabilities among sectors to address local issues (e.g., coastal America); stakeholder focus groups and roundtables that bring together parties with similar vested interests; and governmental advisory groups, such as the Ocean Research Advisory Panel (ORAP).

Marine planning efforts should also provide a significant new level of collaboration and coordination across multiple sectors and levels. Although many of these efforts are initiated at the Federal level, they represent forums for discussion and interaction among different groups.

Mechanisms for international collaboration including international treaties, bilateral agreements, and international scientific forums are also essential for the Nation to more effectively pursue global research efforts that extend outside our borders, both in terms of influence and impact (see International Cooperation, below).

Supporting National Priorities

Working together to protect and sustain-ably use the resources offered by the ocean, our coasts and coastal watersheds, and the Great Lakes will advance other national priorities, including a robust economy, sustainable development, improved human health, national and homeland security, and energy independence (see Box 11). Sustainable development will require the input of multiple agencies and sectors. The ocean is complex and its components complexly linked. None of these goals can be reached independently; all sectors should collaborate to continue moving forward. Mechanisms for integration and collaboration across agencies and private sectors will be essential in meeting national priority objectives.

Box 11: Supporting the National Ocean Policy: Collaboration

The importance of mechanisms for collaboration is highlighted in the National Ocean Policy, which states the need for a "comprehensive framework to coordinate efforts among Federal, State, tribal, and local authorities, including regional governance structures, non-governmental organizations, the private sector, and the public."

The proposed framework includes a means for greater participation and coordination among these groups and requires a process for external input. Accomplishing the National Ocean Policy priorities will require greater coordination and communication across jurisdictional lines.

Beach near the port of Galilee, Point Judith, Rhode Islands. Credit: Mr. William B. Folsom, National Marine Fisheries Service, National Oceanic and Atmospheric Administration.

Making Progress

Although existing mechanisms for coordination should be used wherever possible, new mechanisms may be required to enable effective partnerships and collaboration. It is critical that mechanisms for collaboration at all levels remain flexible in order to address specific needs at the appropriate scale. Requirements include:

- Expansion of interactions with other sectors, promoting collaborations among sectors, and advancing regional initiatives to meet ocean research objectives;
- Development of regional mechanisms for areas where regional associations are not currently in place;
- Establishment of mechanisms to consistently develop broad-based regional research structures that provide effective, ongoing regional interaction;
- Consideration of new mechanisms, including industry-university cooperative research centers that promote long-term partnerships among industry, academic institutions, and government;
- Increased coordination between resource managers and scientists to help ensure that scientific knowledge is translated into management action; and

- Build upon the emphasis on the framework for marine planning in the National Ocean Policy.

International Communication and Cooperation

With a single global ocean bathing the world's shores, international communication and cooperation are intrinsic to both ocean science and policy. Federal agencies partner with many countries on a wide variety of activities and are committed to improving international partnerships on both governmentto-government and scientist-to-scientist levels; the need is particularly strong for the southern hemisphere, where regional science and management collaboration are weakest. International partnerships also broaden the intellectual talent pool and can leverage U.S. investments. International commitment to shared data standards and rapid data access are central to maximizing progress, as are increased coordination and integration of ocean observations. Federal agencies have entered into a variety of international management agreements (e.g., for fisheries management) that require science collaboration.

Overview of Status

Even when nations do not agree politically, very often their scientists continue to cooperate and maintain needed connections. By capitalizing on and expanding international efforts, nations can more effectively pursue global research efforts that extend beyond borders, both in terms of influence and impact. The Intergovernmental Oceanographic Commission coordinates programs in marine research, observation systems, hazard mitigation, and capacity development to provide information to manage ocean and coastal resources more effectively. The Intergovernmental Panel on Climate Change engages thousands of scientists from 194 countries to assess the most current information on climate change and provides rigorous and balanced information to decision makers. The International Council on the Exploration of the Sea focuses its scientific gaze on the North Atlantic and adjacent regional seas. It promotes research on oceanography, marine ecosystems, living marine resources, and human impacts on the ocean and provides advice to its 20 member nations. Its sister organization, the North Pacific Marine Science Organization, has a similar mandate but focuses its attention on the northern North Pacific and adjacent seas. These are just a few examples of international scientific cooperation. Many other organizations, both large and small, foster international partnerships, collaborations, and friendships that further the goals outlined herein. There is a need to create and support such mechanisms to strengthen science in the southern hemisphere as well.

Supporting National Priorities

As populations grow and pressures build, the world continues to shrink. In this landscape political realms are as interconnected as ecological ones. Success in ensuring economic growth and stability, sustainable and ecologically sensitive development, providing secure borders, and achieving human health priorities is dependent upon cooperation and collaboration across national boundaries (see Box 12).

Box 12: Supporting the National Ocean Policy: International Cooperation

The National Ocean Policy calls for exercising rights and jurisdiction and performing duties in accordance with applicable international law, including respect for and preservation of navigational rights and freedoms, which are essential for the global economy and international peace and security. One of the objectives is to improve coordination and integration across the Federal Government, and as appropriate, engage with the international community. International cooperation across scientific frontiers with the exchange of information and ideas provides a means of making connections and pursuing research activities that can help ensure international cooperation in addressing complex management and policy issues.

Beach near the port of Galilee, Point Judith, Rhode Islands. Credit: Mr. William B. Folsom, National Marine Fisheries Service, National Oceanic and Atmospheric Administration.

Making Progress

Progress has been made through a number of international entities, but connectivity and coordination is still needed in several areas:

- Integration in terms of data compatibility. The compatibility of datasets is a critical challenge;
- Connection of observations on the cryosphere, atmosphere, planetary boundary layers, land, and ocean in models along with ecosystems, organisms, and humans over different scales to better understand weather, climate, and ecosystem processes and services. This level of integration requires an interdisciplinary level of thinking beyond what exists today;
- Advocacy for more effective regional implementation of ocean policy objectives through internal collaboration;
- Establishment of international minimum standards for data collection to facilitate data sharing around the globe. Such sharing would require new capabilities to address gaps or differences in data collection, sharing, and interoperability of technologies, and permit integration of existing research into operational models; and
- Education and development of scientists, managers, and other experts capable of operating in this environment and able to integrate physical, ecological, mathematical, and social data into decision-making tools.

ACHIEVING NATIONAL OCEAN POLICY PRIORITIES: PROGRESS THROUGH OCEAN RESEARCH

Charting the Course highlighted four targeted areas for implementation referred to as Near-term Priorities:

- Comparative Analysis of Marine Ecosystem Organization
- Response of Coastal Ecosystems to Persistent Forcing and Extreme Events
- Assessing Meridional Overturning Ocean Circulation Variability" Implications for Rapid Climate Change
- Sensors for Marine Ecosystems

As a result of *Charting the Course*, a number of agencies have provided funding for both the research priorities within the six societal themes, as well as making specific investments in the Near-term Priorities. The following text highlights the targeted investments made in the last three years, and provides examples of their role in supporting the National Ocean Policy and helping to frame forthcoming Strategic Action Plans for the nine policy objectives. The policy objectives are the heart of the path ahead to further the National Ocean Policy; the 20 research priorities described in this document lend the science to achieve them.

Ecosystem Assessments

Ecosystem-based management and regional ecosystem protection and restoration are two objectives highlighted in the National Ocean Policy. They emphasize managing ocean, coastal, coastal watershed, and Great Lakes resources on the scale of the ecosystem rather than according to individual species or activities or on the basis of political jurisdictions. At the heart of these objectives is understanding how ecosystems work and which management practices are most effective. *Charting the Course* called for this sort of basic research, spurring investment in the Comparative Analysis of Marine Ecosystem Organization (CAMEO). CAMEO supports fundamental research to understand the complex dynamics that control ecosystem structure, productivity, behavior, resilience, and population connectivity. It also addresses the effects of climate variability and human pressures on living resources and their critical habitats. Currently, fifteen CAMEO projects and nearly 100 scientists across the Nation are evaluating the impacts of climate change and harvesting strategies on marine ecosystems. Projects are led by teams of Federal and academic scientists and are designed to bridge the gap between fundamental science and management needs. Selected topics include:

- Development of new statistical models to examine trophic interactions in the Bering Sea and Gulf of Alaska under different climate and fishing regimes;
- Comparison of global ecosystems to determine their resilience to fishing pressure;
- Examination of natural and human influences on coral reef community structure, diversity, and resilience;

- Integration of advanced modeling tools to examine the roles of physical oceanography, biology at lower trophic levels, and fish ecology in forage fish populations in the California Current and Oyashio/Kuroshio Current Systems; and
- Development of advanced forecasting tools to predict future states of ecosystems.

CAMEO now includes 16 projects and more than 80 researchers from 25 academic institutions and 10 NOAA laboratories. Seven projects started in 2009 will be completed in the summer of 2011 with some of them extended into 2012. Nine projects started in 2010 will be completed in 2012. No new awards have been made in 2011. NOAA and NSF are working together on upcoming CAMEO workshops and meeting activities, and discussing out-year planning for the program.

Coastal Impacts

The National Ocean Policy provides a framework for marine planning that incorporates the principles of ecosystem-based management. Marine planning is intended to be a comprehensive, transparent public policy process for analyzing current and anticipated uses of ocean, coastal, coastal watershed, and Great Lakes areas to enable society to better determine how they should be sustainably used and protected.

Charting the Course emphasized Forecasting the Response of Coastal Ecosystems to Persistent Forcing and Extreme Events. This effort aims to enhance and integrate observations, monitoring, and modeling to create routinely available information tools that enable better decisions in planning, management, mitigation, and safety. Action has been initiated in two regions on issues of local concern. In the Gulf of Mexico, the effort is focused on integrating diverse regionally-based data and models to address impacts of coastal inundation. In the Great Lakes, the focus is on integrating enhancements of methods, monitoring, and models to improve predictions of beach water quality. Progress has been made through this effort both in furthering research and in developing decision-support tools by

- Supporting research projects to address climate and natural hazard resilience in the Gulf of Mexico;

- Developing decision-support visualization tools for Delaware and the Alabama/Mississippi coasts;
- Developing tools to support sustainable development and site selection for public housing investments; and,
- Enhancing and leveraging relevant sea-level rise and inundation efforts across the U.S. Gulf region.

Recent extreme weather events underscore the importance and timeliness of this effort. The Federal leads, NOAA and the U.S. Geological Survey (USGS), remain committed to dedicating resources toward activities that will advance this near-term priority despite potential budgetary challenges. Charting the Course no longer exists as a line item within NOAA's budget and the USGS budget outlook is uncertain; however, providing research, products, and services to allow coastal zone managers to anticipate and respond to the consequences of coastal change, including storms, inundation, erosion, and sea-level rise, remain high priorities for NOAA and the USGS coastal programs.

Climate Drivers

The National Ocean Policy highlights the impact of changing climate on the ocean and Great Lakes and along our coasts. Changing sea level and water levels in the Great Lakes, warmer water temperatures, decreasing ocean salinity, and increasing ocean acidification will have profound impacts on our shorelines and ecosystems, and the communities and economies that depend on them.

We now realize the importance of ocean circulation in the planet's response to increasing atmospheric CO_2 and warming. An interagency program[53] is addressing the role in climate variability of the ocean circulation feature called the Atlantic Meridional Overturning Circulation, a Near-term Priority in Charting the Course. This effort is working toward developing an AMOC observing system; assessing our ability to reconstruct and model past AMOC changes and predict future changes; and assessing AMOC's impact on regional and global climate, sea-level rise, carbon uptake, and marine ecosystems. Targeted investments in the United States have been strongly leveraged by European investments, in particular connected to a joint project with the United Kingdom. With close ties to U.S. Global Change Research Program activities and international research cooperation, the AMOC program is making strides toward better understanding this influential component of our planet's climate system and its far-reaching impacts, including:

- Determining the most efficient observing system design, examining satellite datasets, and analyzing data products to identify changes in ocean circulation and provide estimates of AMOC variability;
- Continuously monitoring the AMOC and defining emerging patterns and trends;
- Planning key new components of the AMOC observing system;
- Using models to show variability in Atlantic sea-surface temperature linked to AMOC variability, which mirrors a pattern called Atlantic Multidecadal Variability (AMV). Models indicate that AMOC/AMV variability impacts the global water cycle, causing droughts or extreme rainfall, and affecting monsoons;
- Identifying many potential mechanisms that lead to changes in the AMOC (planned model studies will validate these mechanisms and their "fingerprints" in observations); and
- Providing new insights about the state of the AMOC and its recent variations through novel approaches that combine satellite information and *in situ* observations.

The AMOC Science Team continues development of the AMOC Program with annual meetings and strong collaboration with European, South American and South African colleagues.

Sensors for Observations

A policy objective in the National Ocean Policy emphasizes the need to strengthen and integrate Federal and non-Federal ocean infrastructure, sensors, and data collection into a nation- and world-wide system that makes data quickly and widely available. Rapid development of sensor capability is fundamental to progress. As a result of *Charting the Course*, Federal agencies have made significant investments in developing and commercializing sensors for ecosystem observations through the Near-term Priority Sensors for Marine Ecosystems. The objective of this effort is to help answer questions on ocean and ecosystem processes by measuring biological, chemical, biogeochemical, and bio-optical properties, from individual cells to the global ocean. This effort stresses technological innovation, with small businesses working with scientists from academia and government. With multi-agency funding through the National Oceanographic Partnership Program, many competitive and peer-reviewed projects have been funded, with investments put toward developing or commercializing unique tools with which to gather answers about the ocean and its inhabitants. Project results so far include:

- Automated pH sensors showing a lock-step relationship between CO_2 in the atmosphere and the pH of seawater;
- Highly sensitive CO_2 sensors improved to allow for wide-scale ocean monitoring, including float systems for use in dynamic ocean regions;
- An automated analyzer measuring isotopes in methane gas as it emerges from the sea floor;
- Chemical sensors in production for long-term commercial measurement of important ocean food web nutrients; and
- An imaging instrument to view and continuously count individual phytoplankton cells, being transitioned to commercial work.

Projects funded through the Sensors effort continue to provide valuable results. Agencies involved in this effort plan to continue working through the National Oceanographic Partnership Program to fund additional projects in the future.

LOOKING FORWARD

Charting the Course included an Implementation Strategy to guide a nationwide effort to make progress toward the research priorities. The Nation now has a comprehensive, integrated National Ocean Policy for the stewardship of the ocean, our coasts, and the Great Lakes. It sets our Nation on a path toward comprehensive planning for their preservation and sustainable use. The actions in the Implementation Plan for the National Ocean Policy will incorporate the type of scientific data, observations, and collaborative actions that underlie the research priorities highlighted in Science for an Ocean Nation. Scientific discovery, driven by competitive, peer-reviewed investigations, is critical to implementing the National Ocean Policy. Progress requires the continued support of both systematic measurements of the ocean's properties and the freedom to pursue new ideas and technology. Science for an Ocean Nation emphasizes the research efforts with particular anticipated societal applications, while invoking the fundamental research that provides the foundation for those applications.

This document addresses one of the actions included in the National Ocean Policy Implementation Plan. It makes significant progress in identifying priorities in addressing emerging issues and changes in the ocean, our coasts, and the Great Lakes, and highlighting specific scientific requirements and research needs. This report also puts in place many pieces that will be integral to informing implementation of the National Ocean Policy overall. The updated priorities for ocean science and technology in Science for and Ocean Nation cross a wide range of societal interests. Developed with input from many sectors of the ocean community, this document concisely gathers together the research issues recognized as priorities by that community and demonstrates their ties to management needs and policy decisions.

Implementing the National Ocean Policy will require effective cooperation among Federal ocean agencies. In developing and now updating the ocean research priorities, the SOST agencies have developed a strong cooperative network for advancing ocean science and technology, responding to Congress, and coordinating agency activities. This well-developed inclination toward, and mechanism for, collaboration continues in the governance structure established by the National Ocean Policy. A number of Interagency Working Groups contribute to the science, education and technology called for in this document. Some are at work developing research and other plans called for in the Omnibus Public Lands Management Act of 2009 (Public Law 11-11), which will expedite development of the strategic action plans for the National Ocean Policy. Review of this document by the ORAP has broadened the discussion and helped to strengthen the framework for accomplishing the actions necessary to implement the National Ocean Policy.

Charting the Course served the ocean, our coasts, and the Great Lakes well by immediately raising their visibility and focusing efforts to support programs and projects. It also guided priorities in Federal agency budget submissions. Since that report was released, our understanding of the ocean has improved, and we now have our Nation's first National Ocean Policy. Science for an Ocean Nation reflects progress as well as emerging needs, and showcases the ocean science and technology that underpin the goals of the National Ocean Policy.

ABBREVIATIONS AND ACRONYMS

AMOC — Atlantic Meridional Overturning Circulation
AMV — Atlantic Multidecadal Variability
BOEMRE — Bureau of Ocean Energy Management, Regulation and Enforcement
CAMEO — Comparative Analysis of Marine Ecosystem Organization
CMTS — Committee on the Marine Transportation System
CO2 — Carbon Dioxide
COHH — Centers for Oceans and Human Health
COSEE — Centers for Ocean Sciences Education Excellence
EBM — ecosystem-based management
EEZ — Exclusive Economic Zone
FOARAM — Federal Ocean Acidification Research and Monitoring Act
FSV — fisheries survey vessel
GCRP — Global Change Research Program
GDP — gross domestic product
GIS — geographic information systems
HAB — harmful algal blooms
HABISS — Harmful Algal Bloom-related Illness Surveillance System
HEWS — Health Early Warning System
ICOOSA — Integrated Coastal and Ocean Observation System Act
ICES — International Council on the Exploration of the Sea
IEA — Integrated Ecosystem Assessment
IWG-4H — Interagency Working Group on Harmful Algal Blooms, Hypoxia, and Human Health
IPCC — Intergovernmental Panel on Climate Change
IWG-OA — Interagency Working Group on Ocean Acidification
LCC — Landscape Conservation Cooperatives
LME — large marine ecosystem
MMS — Minerals Management Service
MTS — marine transportation system
NAO — North Atlantic Oscillation
NGO — non-governmental organization
NOAA — National Oceanic and Atmospheric Administration
NOC — National Ocean Council
NRC — National Research Council
NTP — near-term priority
OCS — outer continental shelf

OHH	ocean and human health
OHHI	Oceans and Human Health Initiative
OOI	Ocean Observatories Initiative
ORAP	Ocean Research Advisory Panel
PCAST	President's Council of Advisors on Science and Technology
PDO	Pacific Decadal Oscillation
PICES	North Pacific Marine Science Organization
R/V	research vessel
SDR	Subcommittee for Disaster Reduction (of the National Science and Technology Council)
SOST	Subcommittee on Ocean Science and Technology (of the National Science and Technology Council)
STEM	science, technology, engineering, and math
TFUS	Task Force on Unmanned Systems
UNOLS	University National Oceanographic Laboratory Systems

End Notes

[1] National Marine Fisheries Service Office of Science and Technology. 2012. Fisheries in the United States – 2011. http://www.st.nmfs.noaa.gov/Assets/commercial/fus/fus11 /FUS2011. pdf (accessed November 2012).

[2] NOAA stics.noaa.gov estimates of BEA GDP, 2011 (as reported on www.stateofthecoast. noaa.gov, accessed May 2012)

[3] BLS, 2011; World Bank, 2011 (as reported on www.stateofthecoast.noaa.gov, accessed May 2012)

[4] Available at http://www.presidency.ucsb.edu/ws/index.php?pid=88216.

[5] White House Council on Environmental Quality. 2010. Final Recommendations of the Interagency Ocean Policy Task Force. http://www.whitehouse.gov/administration/eop /oceans (accessed October 2010).

[6] Four listening sessions were held at major community conferences in 2009 and 2010. Forty-three sets of comments were received as a result of a Federal Register Notice requesting public input.

[7] National Science and Technology Council, Joint Subcommittee on Ocean Science and Technology. 2007. Charting the Course for Ocean Science in the United States for the Next Decade: An Ocean Research Priorities Plan and Implementation Strategy. http://www.whitehouse.gov/sites/default/files/ microsites/ostp/nstc-orppis.pdf (accessed October 2010).

[8] White House Council on Environmental Quality. 2010. Final Recommendations of the Interagency Ocean Policy Task Force. http://www.whitehouse.gov/administration/eop /oceans (accessed July 2010).

[9] National Oceanographic and Atmospheric Administration. 2010. New NOAA website highlights economic and ecological value of America's coast. http://www.noaanews.noaa. gov /stories2010/20100622_ stateofcoasts.html (accessed July 2010).

[10] National Oceanic and Atmospheric Administration. The Economics and Social Benefits of NOAA Data and Products. http://economics.noaa.gov/?goal=weather&file=events/erosion (accessed December 2010).

[11] John Heinz III Center for Science Economics and the Environment. 2000. The Hidden Costs of Coastal Hazards: Implications for Risk Assessment and Mitigation. Washington, DC: Island Press.

[12] Costanza, R., Perez-Maqueo, O., Martinez, M., Sutton, P., Anderson S., and Mulder, K. 2008. The Value of Coastal Wetlands for Hurricane Protection. AMBIO: A Journal of the Human Environment.

[13] Minerals Management Service. 2006. Technology White Paper on Ocean Current Energy Potential on the U.S. Outer Continental Shelf. http://ocsenergy.anl.gov/documents/docs /OCS_EIS_WhitePaper_ Current.pdf (accessed July 2010).

[14] National Academy of Science, Committee on the Development of an Integrated Science Strategy for Ocean Acidification Monitoring, Research, and Impacts Assessment. 2010. http://books.nap.edu/openbook.php?booksearch=1&term=ocean%20 acidification&record_ id=12904 (accessed December 2010).

[15] U.S. Arctic Research Commission. 2010. Report on Goals and Objectives for Arctic Research 2009– 2010. http://www.arctic.gov/publications/usarc_2009-10_goals.pdf (accessed July 2010).

[16] Bush, President George W. 2009. National Security Presidential Directive – 66/Homeland Security Presidential Directive – 25. http://www.fas.org/irp/offdocs/nspd/nspd-66.htm (accessed July 2010).

[17] Approximately 3.4 million square miles. U.S. Coast Guard 2009. Living Marine Resources. http:// www.uscg.mil/hq/cg5/cg531/LMR.asp (accessed August 2010).

[18] The United Nations Convention on the Law of the Sea allows coastal nations to establish an exclusive economic zone (EEZ) adjacent to its territorial sea, extending a maximum of 200 miles seaward from the baseline. Within its EEZ, the coastal nation has sovereign rights for the purpose of exploring, exploiting, conserving, and managing living and nonliving resources, whether found in ocean waters, the seabed, or subsoil. It also has jurisdiction over artificial islands or other structures with economic purposes. U.S. Commission on Ocean Policy. 2004. An Ocean Blueprint for the 21st Century. Washington, DC,: GPO, p. 35. http://www.oceancommission.gov (accessed July 2010).

[19] National Oceanic and Atmospheric Administration, National Marine Fisheries Service. 2009. Fisheries Economics of the United States 2006. Sociocultural status and trends. http://www.st.nmfs.noaa.gov/st5/ publication/fisheries_economics_2008.html (accessed November 2010).

[20] National Oceanic and Atmospheric Administration, National Marine Fisheries Service. 2009. Fisheries Economics of the United States 2006. Sociocultural status and trends. http://www.st.nmfs.noaa.gov/st5/ publication/fisheries_economics_2008.html (accessed November 2010).

[21] Minerals Management Service. 2009. Survey of Available Data on OCS Resources and Identification of Data Gaps. MMS 2009-015. http://www.boemre.gov/5-year/PDFs/45-DayReportAvailableDataOnOff shoreResources.pdf (accessed July 2010).

[22] Musial, Walter, and Bonnie Ram. 2010. Large-Scale Offshore Wind Power in the United States: Assessment of Opportunities and Barriers Executive Summary. National Renewable Energy Laboratory. http://www.nrel.gov/docs/fy10osti/49229.pdf (accessed October 2010).

[23] Minerals Management Service. 2009. Survey of Available Data on OCS Resources and Identification of Data Gaps. MMS 2009-015. http://www.boemre.gov/5-year/PDFs/45-DayReportAvailableDataOnOff shoreResources.pdf (accessed July 2010).

[24] Collett, T.S. 1995. Gas hydrate resources of the United States. In Gautier, D.L., Dolton, G.L., Takahashi, K.I., and Varnes, K.L., eds., 1995. National Assessment of United States Oil and Gas Resources. CD-ROM. U.S. Geological Survey Digital Data Series 30.

[25] Collett, T.S. 1997. Resource potential of marine and permafrost associated gas hydrates. In Max, M.D., Pallenbarg, R.E., and Rath, B.B., eds., Oceanic Gas Hydrate: Guidance for Research and Programmatic Development at the Naval Research Laboratory. Proceedings of the Workshop on Naval Research Laboratory Gas Hydrate Research Program, September 23–24, 1997. NRL/MR/6100-97-8124, 51 pp.

[26] Great Lakes Environmental Research Laboratory. 2004. About our Great Lakes. http://www.glerl. noaa.gov/pr/ourlakes/ (accessed August 2010).

[27] U.S. Commission on Ocean Policy. 2004. An Ocean Blueprint for the 21st Century. Final Report Washington, DC,: GPO, p. 35. http://www.oceancommission.gov (accessed August 2010).

[28] Crossett, K.M., et al. 2004. Population Trends Along the Coastal United States: 1980–2008. NOAA Coastal Trends Report. GPO: Washington, DC. http://www.oceanservice.noaa.gov /programs/mb/ supp_cstl_population.html (accessed August 2010).

[29] National Science and Technology Council, Subcommittee on Disaster Reduction. 2005. Grand Challenges for Disaster Reduction. http://www.sdr.gov/GrandChallengesSecondPrinting.pdf (accessed July 2010).

[30] U.S. Commission on Ocean Policy. 2004. An Ocean Blueprint for the 21st Century. Final Report. Washington, D.C.: GPO, p.192.

[31] For example, as proposed by the Multipurpose Marine Cadastre Initiative. U.S. Department of Interior, Minerals Management Service. 2005. OCS Mapping Initiative: Energy Policy Act of 2005, Implementation Plan for the Multipurpose Marine Cadastre. http://www.mms.gov/ld/PDFs/MappingInitiative.pdf (accessed August 2010).

[32] U.S. Global Change Research Program. 2009. Our Changing Planet: The U.S. Climate Change Science Program for Fiscal Year 2010. http://downloads.globalchange.gov/ocp/ocp 2010/ocp2010.pdf (accessed July 2010).

[33] U.S. Climate Variability and Predictability Research Program. 2009. Second Annual Progress Report for a JSOST Near-Term Priority Assessing Meridional Overturning Circulation Variability :Implications for Rapid Climate Change. U.S. CLIVAR Office Rep. 2009-1, 25 pp.

[34] McLeod, K.L., and H.M. Leslie, eds. 2009. Ecosystem-Based Management for the Oceans. Washington, DC: Island Press.

[35] Thayer, G.W., T.A. McTigue, R.J. Bellmer, F.M. Burrows, D.H. Merkey, A.D. Nickens, S.J. Lozano, P.F. Gayaldo, P.J. Polmateer, and P.T. Pinit. 2003. Science-Based Restoration Monitoring of Coastal Habitats. Vol. 1, A Framework for Monitoring Plans under the Estuaries and Clean Waters Act of 2000 (Public Law 160-457). NOAA Coastal Ocean Program Decision Analysis no. 23. NOAA National Centers for Coastal Ocean Science, Silver Spring, MD. 35 pp. plus appendices.

[36] Thayer, G.W., T.A. McTigue, R.J. Salz, D.H. Merkey, F.M. Burrows, and P.F. Gayaldo, eds. 2005. Science-Based Restoration Monitoring of Coastal Habitats. Vol. 2, Tools for Monitoring Coastal Habitats. NOAA Coastal Ocean Program Decision Analysis no. 23. NOAA National Centers for Coastal Ocean Science, Silver Spring, MD. 628 pp. plus appendices.

[37] The study of evolutionary relationships of species.

[38] Millennium Ecosystem Assessment. 2005. Ecosystems and Human Well-being: Synthesis. Washington, DC: Island Press. http://www.millenniumassessment.org/documents /document. 356.aspx.pdf (accessed July 2010).

[39] Van Dolah, F.M. 2000. Marine Algal Toxins: Origins, Health Effects, and Their Increased Occurrence. Environ Health Perspect 2000, 108(Suppl 1):133-141.

[40] Lopez, C.B., Dortch, Q., Jewett, E.B., Garrison, D. 2008. Scientific Assessment of Marine Harmful Algal Blooms. Interagency Working Group on Harmful Algal Blooms, Hypoxia, and Human Health of the Joint Subcommittee on Ocean Science and Technology. Washington, D.C.

[41] U.S. Commission on Ocean Policy. 2004. An Ocean Blueprint for the 21st Century. Final Report. Washington, DC: GPO, pp.340–342.

[42] Mozaffarian, D. and E.B. Rimm. 2006. Fish intake, contaminants and health. Evaluating the risks and benefits. Journal of the American Medical Association 296:1,885–1,899.

[43] Bioinformatics is research, development, or application of computational tools and approaches for expanding the use of biological, medical, behavioral or health data, including those to acquire, store, organize, archive, analyze, or visualize such data. National Institutes of Health working definition of bioinformatics. 2000. http://www.bisti.nih.gov/docs /compubiodef.pdf (accessed November 2010).

[44] Global Climate Change Impacts in the United States, Thomas R. Karl, Jerry M. Melillo, and Thomas C. Peterson, (eds.). Cambridge University Press, 2009.

[45] PCAST (2011). Sustaining Environmental Capital: Protecting Society and the Economy. Retrieved from http://www.whitehouse.gov/sites/default/files/microsites/ostp/pcast_ sustaining_environmental_capital_report.pdf

[46] The Ocean Project. 2009. America, the Ocean, and Climate Change: New Research Insights for Conservation, Awareness, and Action. http://www.theoceanproject.org/resources /latest.php (accessed July 2010).

[47] National Geographic Society. 2006. Ocean Literacy: The Essential Principles of Ocean Sciences, K-12. http://oceanservice.noaa.gov/education/literacy/ocean_literacy.pdf (accessed December 2010).

[48] Marine Advanced Technology Education Center. 2009. The Ocean Science, Technology, and Operations Workforce. Presentation to the Interagency Working Group for Ocean Education, Joint Subcommittee on Ocean Science and Technology, Oct. 2009.

[49] American Geological Institute. 2010. Dynamics of the Geoscience Workforce. Presentation to the Interagency Working Group for Ocean Education, Joint Subcommittee on Ocean Science and Technology, Mar. 2010.

[50] Committee on an Ocean Infrastructure Strategy for U.S. Ocean Research in 2030, National Research Council. 2011. Critical Infrastructure for Ocean Research and Societal Needs in 2030. National Academy Press. http://www.nap.edu/catalog.php?record_id=13081 (accessed August 2011).

[51] The National Science and Technology Council's Subcommittee on Ocean Science and Technology is filling the role of the OST-IPC.

[52] See http://www.atlanticmoc.org for more information.

In: The United States as an Ocean Nation ISBN: 978-1-62618-705-4
Editor: Mikkel Jorgensen

Chapter 2

EXECUTIVE ORDER 13547--STEWARDSHIP OF THE OCEAN, OUR COASTS, AND THE GREAT LAKES*

The White House Office of the Press Secretary

By the authority vested in me as President by the Constitution and the laws of the United States of America, it is hereby ordered as follows:

Section 1. Purpose. The ocean, our coasts, and the Great Lakes provide jobs, food, energy resources, ecological services, recreation, and tourism opportunities, and play critical roles in our Nation's transportation, economy, and trade, as well as the global mobility of our Armed Forces and the maintenance of international peace and security. The Deepwater Horizon oil spill in the Gulf of Mexico and resulting environmental crisis is a stark reminder of how vulnerable our marine environments are, and how much communities and the Nation rely on healthy and resilient ocean and coastal ecosystems. America's stewardship of the ocean, our coasts, and the Great Lakes is intrinsically linked to environmental sustainability, human health and well-being, national prosperity, adaptation to climate and other environmental changes, social justice, international diplomacy, and national and homeland security. This order adopts the recommendations of the Interagency Ocean Policy Task Force, except where otherwise provided in this order, and directs executive agencies to implement those recommendations under the guidance

* This Executive Order, authored by President Barack Obama, was released by the White House, Office of the Press Secretary on July19, 2010.

of a National Ocean Council. Based on those recommendations, this order establishes a national policy to ensure the protection, maintenance, and restoration of the health of ocean, coastal, and Great Lakes ecosystems and resources, enhance the sustainability of ocean and coastal economies, preserve our maritime heritage, support sustainable uses and access, provide for adaptive management to enhance our understanding of and capacity to respond to climate change and ocean acidification, and coordinate with our national security and foreign policy interests. This order also provides for the development of coastal and marine spatial plans that build upon and improve existing Federal, State, tribal, local, and regional decision making and planning processes. These regional plans will enable a more integrated, comprehensive, ecosystem-based, flexible, and proactive approach to planning and managing sustainable multiple uses across sectors and improve the conservation of the ocean, our coasts, and the Great Lakes.

Sec. 2. Policy. (a) To achieve an America whose stewardship ensures that the ocean, our coasts, and the Great Lakes are healthy and resilient, safe and productive, and understood and treasured so as to promote the well-being, prosperity, and security of present and future generations, it is the policy of the United States to:

- protect, maintain, and restore the health and biological diversity of ocean, coastal, and Great Lakes ecosystems and resources;
- improve the resiliency of ocean, coastal, and Great Lakes ecosystems, communities, and economies;
- bolster the conservation and sustainable uses of land in ways that will improve the health of ocean, coastal, and Great Lakes ecosystems;
- use the best available science and knowledge to inform decisions affecting the ocean, our coasts, and the Great Lakes, and enhance humanity's capacity to understand, respond, and adapt to a changing global environment;
- support sustainable, safe, secure, and productive access to, and uses of the ocean, our coasts, and the Great Lakes;
- respect and preserve our Nation's maritime heritage, including our social, cultural, recreational, and historical values;
- exercise rights and jurisdiction and perform duties in accordance with applicable international law, including respect for and preservation of navigational rights and freedoms, which are essential for the global economy and international peace and security;

- increase scientific understanding of ocean, coastal, and Great Lakes ecosystems as part of the global interconnected systems of air, land, ice, and water, including their relationships to humans and their activities;
- improve our understanding and awareness of changing environmental conditions, trends, and their causes, and of human activities taking place in ocean, coastal, and Great Lakes waters; and
- foster a public understanding of the value of the ocean, our coasts, and the Great Lakes to build a foundation for improved stewardship.

(b) The United States shall promote this policy by:

- ensuring a comprehensive and collaborative framework for the stewardship of the ocean, our coasts, and the Great Lakes that facilitates cohesive actions across the Federal Government, as well as participation of State, tribal, and local authorities, regional governance structures, nongovernmental organizations, the public, and the private sector;
- cooperating and exercising leadership at the international level;
- pursuing the United States' accession to the Law of the Sea Convention; and
- supporting ocean stewardship in a fiscally responsible manner.

Sec. 3. Definitions. As used in this order:

a) "Final Recommendations" means the Final Recommendations of the Interagency Ocean Policy Task Force that shall be made publicly available and for which a notice of public availability shall be published in the Federal Register.

b) The term "coastal and marine spatial planning" means a comprehensive, adaptive, integrated, ecosystem-based, and transparent spatial planning process, based on sound science, for analyzing current and anticipated uses of ocean, coastal, and Great Lakes areas. Coastal and marine spatial planning identifies areas most suitable for various types or classes of activities in order to reduce conflicts among uses, reduce environmental impacts, facilitate compatible uses, and preserve critical ecosystem services to meet economic, environmental, security, and social objectives. In practical terms, coastal and marine spatial planning provides a public policy process for society to better determine how the ocean, our coasts, and Great Lakes are sustainably used and protected -- now and for future generations.

c) The term "coastal and marine spatial plans" means the plans that are certified by the National Ocean Council as developed in accordance with the definition, goals, principles, and process described in the Final Recommendations.

Sec. 4. Establishment of National Ocean Council. (a) There is hereby established the National Ocean Council (Council).

(b) The Council shall consist of the following:

- the Chair of the Council on Environmental Quality and the Director of the Office of Science and Technology Policy, who shall be the Co-Chairs of the Council;
- the Secretaries of State, Defense, the Interior, Agriculture, Health and Human Services, Commerce, Labor, Transportation, Energy, and Homeland Security, the Attorney General, the Administrator of the Environmental Protection Agency, the Director of the Office of Management and Budget, the Under Secretary of Commerce for Oceans and Atmosphere (Administrator of the National Oceanic and Atmospheric Administration), the Administrator of the National Aeronautics and Space Administration, the Director of National Intelligence, the Director of the National Science Foundation, and the Chairman of the Joint Chiefs of Staff;
- the National Security Advisor and the Assistants to the President for Homeland Security and Counterterrorism, Domestic Policy, Energy and Climate Change, and Economic Policy;
- an employee of the Federal Government designated by the Vice President; and
- such other officers or employees of the Federal Government as the Co-Chairs of the Council may from time to time designate.

(c) The Co-Chairs shall invite the participation of the Chairman of the Federal Energy Regulatory Commission, to the extent consistent with the Commission's statutory authorities and legal obligations, and may invite the participation of such other independent agencies as the Council deems appropriate.

(d) The Co-Chairs of the Council, in consultation with the National Security Advisor and the Assistant to the President for Homeland Security and Counterterrorism, shall regularly convene and preside at meetings of the Council, determine its agenda, direct its work, and, as appropriate to address particular subject matters, establish and direct committees of the Council that shall consist exclusively of members of the Council.

(e) A member of the Council may designate, to perform committee functions of the member, any person who is within such member's department, agency, or office and who is (i) an officer of the United States appointed by the President, (ii) a member of the Senior Executive Service or the Senior Intelligence Service, (iii) a general officer or flag officer, or (iv) an employee of the Vice President.

(f) Consistent with applicable law and subject to the availability of appropriations, the Office of Science and Technology Policy and the Council on Environmental Quality shall provide the Council with funding, including through the National Science and Technology Council or the Office of Environmental Quality. The Council on Environmental Quality shall, to the extent permitted by law and subject to the availability of appropriations, provide administrative support necessary to implement this order.

(g) The day-to-day operations of the Council shall be administered by a Director and a Deputy Director, who shall supervise a full-time staff to assist the Co-Chairs in their implementation of this order.

Sec. 5. Functions of the Council. (a) The Council shall have the structure and function and operate as defined in the Final Recommendations. The Council is authorized, after the Council's first year of operation, to make modifications to its structure, function, and operations to improve its effectiveness and efficiency in furthering the policy set forth in section 2 of this order.

(b) To implement the policy set forth in section 2 of this order, the Council shall provide appropriate direction to ensure that executive departments', agencies', or offices' decisions and actions affecting the ocean, our coasts, and the Great Lakes will be guided by the stewardship principles and national priority objectives set forth in the Final Recommendations, to the extent consistent with applicable law. The Council shall base its decisions on the consensus of its members. With respect to those matters in which consensus cannot be reached, the National Security Advisor shall coordinate with the Co-Chairs and, as appropriate, the Assistants to the President for Energy and Climate Change, and Economic Policy, and the employee of the United States designated by the Vice President, subject to the limitations set forth in section 9 of this order, to present the disputed issue or issues for decision by the President.

Sec. 6. Agency Responsibilities. (a) All executive departments, agencies, and offices that are members of the Council and any other executive department, agency, or office whose actions affect the ocean, our coasts, and the Great Lakes shall, to the fullest extent consistent with applicable law:

- take such action as necessary to implement the policy set forth in section 2 of this order and the stewardship principles and national priority objectives as set forth in the Final Recommendations and subsequent guidance from the Council; and
- participate in the process for coastal and marine spatial planning and comply with Council certified coastal and marine spatial plans, as described in the Final Recommendations and subsequent guidance from the Council.

(b) Each executive department, agency, and office that is required to take actions under this order shall prepare and make publicly available an annual report including a concise description of actions taken by the agency in the previous calendar year to implement the order, a description of written comments by persons or organizations regarding the agency's compliance with this order, and the agency's response to such comments.

(c) Each executive department, agency, and office that is required to take actions under this order shall coordinate and contribute resources, as appropriate, to assist in establishing a common information management system as defined in the Final Recommendations and shall be held accountable for managing its own information assets by keeping them current, easily accessible, and consistent with Federal standards.

(d) To the extent permitted by law, executive departments, agencies, and offices shall provide the Council such information, support, and assistance as the Council, through the Co-Chairs, may request.

Sec. 7. Governance Coordinating Committee. The Council shall establish a Governance Coordinating Committee that shall consist of 18 officials from State, tribal, and local governments in accordance with the Final Recommendations. The Committee may establish subcommittees chaired by representatives of the Governance Coordinating Committee. These subcommittees may include additional representatives from State, tribal, and local governments, as appropriate to provide for greater collaboration and diversity of views.

Sec. 8. Regional Advisory Committees. The lead Federal department, agency, or office for each regional planning body established for the development of regional coastal and marine spatial plans, in consultation with their nonfederal co-lead agencies and membership of their regional planning body, shall establish such advisory committees under the Federal Advisory Committee Act, 5 U.S.C. App., as they deem necessary to provide information and to advise the regional planning body on the development of regional coastal and marine spatial plans to promote the policy established in section 2 of this order.

Sec. 9. General Provisions. (a) Nothing in this order, the establishment of the Council, and the Final Recommendations shall be construed to impair or otherwise affect:

- authority granted by law to an executive department or agency or the head thereof; or
- functions assigned by the President to the National Security Council or Homeland Security Council (including subordinate bodies) relating to matters affecting foreign affairs, national security, homeland security, or intelligence.

(b) Nothing in this order shall be construed to impair or otherwise affect the functions of the Director of the Office of Management and Budget relating to budgetary, administrative, or legislative proposals.

(c) In carrying out the provisions of this order and implementing the Final Recommendations, all actions of the Council and the executive departments, agencies, and offices that constitute it shall be consistent with applicable international law, including customary international law, such as that reflected in the Law of the Sea Convention.

(d) This order is not intended to, and does not, create any right or benefit, substantive or procedural, enforceable at law or in equity by any party against the United States, its departments, agencies, or entities, its officers, employees, or agents, or any other person.

Sec. 10. Revocation. Executive Order 13366 of December 17, 2004, is hereby revoked.

BARACK OBAMA

THE WHITE HOUSE, July 19, 2010.

INDEX

A

B

C

D

E

F

G

H

N

O

P

Q

R

S

W